T0178603

SpringerBriefs in Mathematics

SpringerBriefs in Mathematics showcases expositions in all areas of mathematics and applied mathematics. Manuscripts presenting new results or a single new result in a classical field, new field, or an emerging topic, applications, or bridges between new results and already published works, are encouraged. The series is intended for mathematicians and applied mathematicians.

More information about this series at http://www.springer.com/series/10030

Kunihiko Kodaira

Nevanlinna Theory

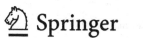
Springer

Author
Kunihiko Kodaira (1915–1997)
The University of Tokyo
Tokyo
Japan

Translated by
Takeo Ohsawa
Graduate School of Mathematics
Nagoya University
Nagoya
Japan

ISSN 2191-8198   ISSN 2191-8201   (electronic)
SpringerBriefs in Mathematics
ISBN 978-981-10-6786-0   ISBN 978-981-10-6787-7   (eBook)
https://doi.org/10.1007/978-981-10-6787-7

Library of Congress Control Number: 2017955244

Mathematics Subject Classification (2010): 32A22, 32H25, 32H30

Printed on acid-free paper

This Springer imprint is published by Springer Nature
The registered company is Springer Nature Singapore Pte Ltd.
The registered company address is: 152 Beach Road, #21-01/04 Gateway East, Singapore 189721, Singapore

Notes taken by Fumio Sakai

Kunihiko Kodaira at his home in Tokyo, 1990 © Springer Japan

Preface

Recently, by the works of Shoshichi Kobayashi, Takushiro Ochiai, J. Carlson, and P. Griffiths, some of the basic results in complex analysis of one complex variable, such as Schwarz lemma and Nevanlinna theory, have been brought into a new scope and successfully extended to several variables in a very elegant manner. The purpose of this lecture is to give an account of such results. The course was planned for one semester at first, but actually it had to be given separately in two semesters, with a break in-between for half a year. As a result, to the author's regret the exposition became not so well organized.

Nowadays, a tendency prevails to regard the newest results as the most important research sources. However, a glance at those results presented here will give an immediate impression that the contrary is also true in some cases. In fact, except for the elementary prerequisites on complex manifolds, the only background needed here is Nevanlinna's monograph which was published in 1936.

The author expresses his deep gratitude to Mr. Fumio Sakai who dictated the lecture by completing the missing details.

(Chapters 1 and 2 were lectured from October 1972 through March 1973 and Chaps. 3 and 4 from October 1973 through March 1974.)

Princeton, USA   Kunihiko Kodaira
March 1974

Contents

Chapter 1
Nevanlinna Theory of One Variable (1)

Abstract The classical Nevanlinna theory is reviewed, after preparation of necessary terminologies in differential geometry and integral formulas.

Keywords Riemann surface · Kähler metric · Gaussian curvature · The first main theorem · Defect relation

1 Metrics on Compact Riemann Surfaces

Let W be a compact Riemann surface and let $W = \bigcup_j U_j$ be an open covering with local coordinates w_j representing $w \in U_j$.

Definition. *A Kähler metric ds^2 on W is*

$$ds^2 = a_j(w)\,|dw_j|^2.$$

Here $a_j(w)$ is a C^∞ function on U_j satisfying $a_j(w) > 0$ and

$$a_j(w) = \left|\frac{dw_k}{dw_j}\right|^2 a_k(w)$$

on $U_j \cap U_k \neq \varnothing$.

The Kähler form ω associated to ds^2 is defined as

$$\omega = \frac{\sqrt{-1}}{2\pi} a_j(w)\,dw_j \wedge d\overline{w_j} \quad \text{(on } U_j\text{)}.$$

© The Author(s) 2017
K. Kodaira, *Nevanlinna Theory*, SpringerBriefs in Mathematics,
https://doi.org/10.1007/978-981-10-6787-7_1

Definition. *The Gaussian curvature of the metric ds^2 is*

$$R(w) = -\frac{2}{a_j(w)} \frac{\partial^2 \log a_j(w)}{\partial w_j \partial \overline{w}_j}.$$

(1.1)*Remark.*

$$\frac{\sqrt{-1}}{2\pi} \partial \overline{\partial} \log a_j(w) = -\frac{R(w)}{2} \omega,$$

where the differentiation with respect to $z = x + \sqrt{-1}y$ is defined by

$$\partial u = \frac{\partial u}{\partial z} dz, \quad \frac{\partial u}{\partial z} = \frac{1}{2} \left(\frac{\partial u}{\partial x} - \sqrt{-1}\frac{\partial u}{\partial y} \right)$$

$$\overline{\partial} u = \frac{\partial u}{\partial \overline{z}} d\overline{z}, \quad \frac{\partial u}{\partial \overline{z}} = \frac{1}{2} \left(\frac{\partial u}{\partial x} + \sqrt{-1}\frac{\partial u}{\partial y} \right)$$

for any differentiable function u in x, y.

Our canonical choices of metrics on W are as follows, where $g = g(W)$ denotes the genus of W.

(i) $g = 0$. If W is the Riemann sphere $\mathbb{P}^1 = \mathbb{C} \cup \{\infty\}$,

$$ds^2 = \frac{|dw|^2}{(1 + |w|^2)^2}, \quad \omega = \frac{\sqrt{-1}}{2\pi} \frac{dw \wedge d\overline{w}}{(1 + |w|^2)^2}, \quad R = 4.$$

(ii) $g = 1$. If W is a torus (i.e. an elliptic curve) $= \mathbb{C}/G$, where $G = \{g \mid g : w \mapsto w + \sum_{j=1}^{2} n_j w_j, n_j \in \mathbb{Z}\}$ for $w_1, w_2 \in \mathbb{C}$ which are linearly independent over $\mathbb{R}$, with respect to the coordinate w of $\mathbb{C}$ one has

$$ds^2 = |dw|^2, \quad \omega = \frac{\sqrt{-1}}{2\pi} dw \wedge d\overline{w}, \quad R = 0.$$

(iii) $g \geq 2$. If W is a general Riemann surface $= \mathbb{D}/G$;
$\mathbb{D} = \{w \in \mathbb{C} \mid |w| < 1\}$, where G is a subgroup of the group of analytic automorphisms of $\mathbb{D}$,

$$ds^2 = \frac{|dw|^2}{(1 - |w|^2)^2}, \quad \omega = \frac{\sqrt{-1}}{2\pi} \frac{dw \wedge d\overline{w}}{(1 - |w|^2)^2}, \quad R = -4.$$

Calculation of the curvature R is as follows. Write ds^2 as

$$ds^2 = a \, |dw|^2, \quad a = \frac{1}{(1 + \sigma |w|^2)^2},$$

where $\sigma = 1, 0, -1$ according to $g = 0, 1, \geq 2$. Then

$$\frac{\partial \log a}{\partial \overline{w}} = -2\frac{\partial \log(1 + \sigma|w|^2)}{\partial \overline{w}} = -\frac{2\sigma w}{1 + \sigma|w|^2},$$

$$\frac{\partial^2 \log a}{\partial w \partial \overline{w}} = -\frac{2\sigma}{1 + \sigma|w|^2} + \frac{2|w|^2\sigma^2}{(1 + \sigma|w|^2)^2} = \frac{-2\sigma}{(1 + \sigma|w|^2)^2} = -2\sigma a,$$

so that $R = 4\sigma$. Thus the evaluations of R are as above, according to $\sigma = \pm 1, 0$.

2   Integral Formulas

Notation

$$\Delta_r : \text{the } r\text{-disc} = \{z \in \mathbb{C} \mid |z| < r\}$$
$$\partial \Delta_r : \text{ the boundary of } \Delta_r = \{z \in \mathbb{C} \mid |z| = r\}.$$

Let W be a compact Riemann surface, let $0 < r_\infty \le \infty$ and let $f : \Delta_{r_\infty} \to W$ be a holomorphic map.

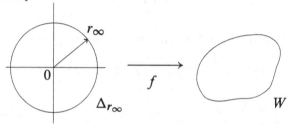

For any point $z \in f^{-1}(U_j)$, we denote $f(z)$ by $f_j(z)$ in terms of the local coordinate w_j, i.e. $f : z \to w_j = f_j(z)$. Then, the pull-back $f^*\omega$ of the Kähler form ω of any metric ds^2 on W is expressed on $f^{-1}(U_j)$ as

$$f^*\omega = \frac{\sqrt{-1}}{2\pi} a_j(f(z))|f_j'(z)|^2 \, dz \wedge d\overline{z}.$$

We put

$$\xi(z) = \frac{1}{\pi} a_j(f(z))|f_j'(z)|^2 \quad \text{on } f^{-1}(U_j),$$

so that $\xi(z)$ is a C^∞ function on Δ_{r_∞} satisfying $\xi(z) \ge 0$. Note that $\xi(z) = 0$ if and only if $f_j'(z) = 0$.

Definition. *For any nonconstant f the roots of $\xi(z) = 0$ are denoted by ρ_h, $h = 1, 2, \ldots$.*

By the above remark, ρ_h are the roots of $f_j'(z) = 0$ on $f^{-1}(U_j)$. By m_h we shall denote the multiplicity of ρ_h.

Definition. (2.1) $M(r) = \frac{1}{2\pi} \int_0^{2\pi} \frac{1}{4\pi} \log \xi(re^{i\theta}) \, d\theta.$

Lemma. $M(r)$ *is piecewise smooth* $(= C^\infty)$ *with respect to* r $(0 \le r < r_\infty)$.

Proof. We shall verify the assertion at $r \ne 0$. For that, let us choose r_1 in such a way that $r < r_1 < r_\infty$ and $r_1 \ne |\rho_h|,\ h = 1, 2, \ldots.$
 Note that

$$f_j'(z) = (z - \rho_h)^{m_h} g_j(z)$$

if $\rho_h \in f^{-1}(U_j)$, where $g_j(z) \ne 0$ on a sufficiently small neighborhood of ρ_h. Hence $\xi(z)$ can be expressed as

$$\xi(z) = \eta(z)|z - \rho_h|^{2m_h},$$

where $\eta_h(z) > 0$ on a sufficiently small neighborhood of ρ_h. Therefore

$$\xi(z) = \eta(z) \prod_{|\rho_h| < r_1} |z - \rho_h|^{2m_h}$$

holds on Δ_{r_1}, where $\eta(z)$ is a C^∞ function on Δ_{r_1} satisfying $\eta(z) > 0$.
 Hence

$$\log \xi(re^{i\theta}) = \log \eta + \sum_{|\rho_h| < r_1} 2m_h \log |re^{i\theta - \rho_h}|$$

so that

$$\int_0^{2\pi} \log \xi(re^{i\theta}) \, d\theta = \int_0^{2\pi} \log \eta \, d\theta + \sum_{|\rho_h| < r_1} 2m_h \int_0^{2\pi} \log |re^{i\theta} - \rho_h| \, d\theta$$

$$= C^\infty \text{ function} + \sum_{|\rho_h| < r_1} 4m_h \pi \max \{\log r, \log |\rho_h|\}.$$

Since $\max \{\log r, \log |\rho_h|\}$ are piecewise smooth, so is $M(r)$ on $[0, r_\infty)$.

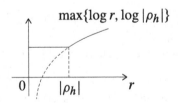

Lemma. (2.2) $-\frac{R}{2} f^* \omega = \frac{\sqrt{-1}}{2\pi} \partial \bar{\partial} \log \xi.$

Proof.

$$\frac{\sqrt{-1}}{2\pi}\partial\bar{\partial}\log\xi = \frac{\sqrt{-1}}{2\pi}\partial\bar{\partial}\log a_j(f_j(z))|f_j'(z)|^2$$

$$= \frac{\sqrt{-1}}{2\pi}\partial\bar{\partial}\{\log a_j(f_j(z)) + \log f_j'(z) + \log\overline{f_j'(z)}\} = \frac{\sqrt{-1}}{2\pi}\partial\bar{\partial}\log a_j(f(z))$$

$$= -\frac{R}{2}f^*\omega.$$

(by Remark (1.1)).

Definition. $d^{\perp} = \sqrt{-1}(\bar{\partial} - \partial)$.

Note that $d^{\perp}u = \frac{\partial u}{\partial x}\,dy - \frac{\partial u}{\partial y}\,dx$ and

$$dd^{\perp}u = 2\sqrt{-1}\partial\bar{\partial}u = \left(\frac{\partial^2 u}{\partial x^2} + \frac{\partial^2 u}{\partial y^2}\right)dx \wedge dy.$$

Let Γ be a C^1-curve in the z-plane parametrized by $(x(s), y(s))$. Here s is chosen to be the arc-length on Γ.

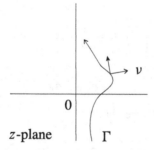

z-plane Γ

Putting

$$\dot{x} = \frac{dx}{ds},\quad \dot{y} = \frac{dy}{ds}\ \text{ and }\ \nu = (\dot{y}, -\dot{x}),$$

one has

$$\int_{\Gamma} d^{\perp}u = \int_{\Gamma}\left(\frac{\partial u}{\partial x}\dot{y} - \frac{\partial u}{\partial y}\dot{x}\right)ds = \int_{\Gamma}\frac{\partial u}{\partial\nu}\,ds,$$

ν being the unit normal vector of Γ and $\frac{\partial u}{\partial\nu}$ denotes the normal derivative. By integrating both sides of Lemma (2.2) one has

$$-\frac{1}{2}\int_{\Delta_r}Rf^*\omega = \frac{1}{4\pi}\int_{\Delta_r}dd^{\perp}\log\xi$$

$$= \frac{1}{4\pi}\int_{\partial\Delta_r}d^{\perp}\log\xi - \sum_{|\rho_h|<r}\frac{1}{4\pi}\oint_{\rho_h}d^{\perp}\log\xi$$

(by Stokes's theorem). Here $\oint_{\rho_h}$ stands for the contour integral along $|z - \rho_h| = \varepsilon$ for sufficiently small ε.

Since $\xi(z) = \eta = h(z)|z - \rho_h|^{2m_h}$ $(\eta_h > 0)$ holds on a neighborhood of ρ_h, one has

$$\frac{1}{4\pi} \oint_{\rho_h} d^{\perp} \log \xi$$
$$= \frac{1}{4\pi} \oint_{\rho_h} \frac{\partial \log \eta_h}{\partial \nu} ds + \frac{m_h}{2\pi} \oint_{\rho_h} \frac{\partial \log |z - \rho_h|}{\partial \nu} ds.$$

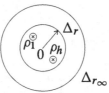

Since the first term on the right hand side is dominated by $(2\pi\varepsilon) \times$ const., it tends to 0 as $\varepsilon \to 0$. The second term is equal to $\sum_{|\rho_h| < r} m_h$ because $\frac{\partial \log |z - \rho_h|}{\partial \nu} = \frac{1}{\varepsilon}$.

Furthermore

$$\frac{1}{4\pi} \int_{\partial \Delta_r} d^{\perp} \log \xi(z) = \frac{1}{4\pi} \int_{\partial \Delta_r} \frac{\partial \log \xi}{\partial \nu} ds$$
$$= \frac{1}{4\pi} \int_0^{2\pi} \frac{\partial \log \xi(re^{i\theta})}{\partial r} r \, d\theta = r \frac{d}{dr} \int_0^{2\pi} \frac{1}{4\pi} \log \xi(re^{i\theta}) \, d\theta = 2\pi r \frac{d}{dr} M(r).$$

Combining these equalities we obtain

(2.3)
$$-\frac{1}{2} \int_{\Delta_r} Rf^*\omega = 2\pi r \frac{d}{dr} M(r) - \sum_{|\rho_h|} m_h.$$

Notation

$$\begin{cases} n_1(r) = \sum_{|\rho_h| < r} m_h = \text{no. of roots of } \xi = 0 \text{ counted wtih multiplicity.} \\ N_1(r) = \int_0^r n_1(t) \frac{dt}{2\pi t} \quad (< \infty \text{ if } \xi(0) \neq 0)). \\ B(r) = -\frac{1}{2} \int_{\delta_r} Rf^*\omega = \int_{\Delta_r} f^*(\frac{\sqrt{-1}}{2\pi} \partial\bar{\partial} \log a_j). \end{cases}$$

Using these, (2.3) is written as

$$\frac{B(r)}{2\pi r} + \frac{n_1(r)}{2\pi r} = \frac{dM(r)}{dr}.$$

Hence, by integrating further we obtain:

Theorem. (2.4) *Under the assumption $\xi(0) \neq 0$ one has*

$$\int_0^r B(t) \frac{dt}{2\pi t} + N_1(t) = M(r) - M(0).$$

3 Holomorphic Maps to Compact Riemann Surfaces of Genus ≥ 2

Let W be a compact Riemann surface of genus ≥ 2 equipped with a canonical metric (cf. Sect. 1, (iii)) whose Kähler form is

$$\omega = \frac{\sqrt{-1}}{2\pi} \frac{dw \wedge d\overline{w}}{(1 - |w|^2)^2}.$$

Theorem. (3.1) *For any holomorphic map $f : \Delta_{r_\infty} \to W$ satisfying $f_j'(0) = 0$,*

$$r_\infty \leq \frac{1}{\sqrt{a_j(f(0))} |f_j'(0)|}$$

holds.

Corollary. *Every holomorphic map $f : \mathbb{C} \to W$ is a constant map.*

The first proof of Theorem (3.1) (by the Schwarz lemma): Since the universal covering $\tilde{W}$ of W is $\mathbb{D}$, there exists a holomorphic map $\tilde{f} : \Delta_{r_\infty} \to \mathbb{D}$ for which the following diagram commutes:

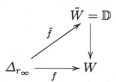

Composing $\tilde{f}$ with an analytic automorphism of $\mathbb{D}$ if necessary, we may assume that $\tilde{f}(0) = 0$. In this situation, since $a_j(0) = \frac{1}{1-|0|^2} = 1$, to prove the assertion it suffices to show that

$$r_\infty \leq \frac{1}{|f_j'(0)|}.$$

By applying the Schwarz lemma (cf. (6.1)) to $\tilde{f}$, one has

$$|\tilde{f}'(0)| \leq \frac{1}{r_\infty}.$$

Since the coordinate w of $\mathbb{D}$ can be taken as the local coordinate w_j, $\tilde{f}$ can be identified with the local expression f_j of f. Consequently

$$r_\infty \leq \frac{1}{|\tilde{f}'(0)|} = \frac{1}{|f_j'(0)|}.$$

The second proof of Theorem (3.1) (by Theorem (2.4)). The proof is actually for the slightly weaker inequality

$$r_\infty \leq \frac{2}{\sqrt{a_j(f(0))}|f_j'(0)|}.$$

First we note that $R = -4$, for the genus of W is ≥ 2. Hence

$$B(r) = 2\int_{\Delta_r} f^*\omega = 2\int_{\Delta_r} \xi(z)\frac{\sqrt{-1}}{2}\,dz\wedge d\bar{z} \quad = 2\int_0^r t\,dt\int_0^{2\pi}\xi(te^{i\theta})\,d\theta.$$

Definition. $\Phi(t) = \frac{1}{2\pi}\int_0^{2\pi}\xi(te^{i\theta})\,d\theta.$

Accordingly

$$B(r) = 4\pi\int_0^r \Phi(t)t\,dt.$$

Next, by the concavity of the logarithm, one has

the average of $\log m(\theta) \leq \log\{\text{the average of }m(\theta)\}$,

so that

$$M(r) = \frac{1}{2\pi}\int_0^{2\pi}\frac{1}{4\pi}\log\xi(rd^{i\theta})\,d\theta \leq \frac{1}{4\pi}\log\left(\frac{1}{2\pi}\int_0^{2\pi}\xi(re^{i\theta})\,d\theta\right)$$

$$= \frac{1}{4\pi}\log\Phi(r).$$

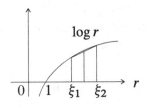

Definition. $Q(r) = \int_0^r B(t) \frac{dt}{2\pi t}$.

Combining the above inequality with Theorem (2.4) we obtain

(3.2)
$$Q(r) \leq \frac{1}{4\pi} \log \Phi(r) - M(0).$$

Now, with respect to another coordinate $\hat{z} = cz$,

$$\frac{df}{d\hat{z}} = \frac{1}{c} \frac{df}{dz}(0).$$

So, by putting $\hat{r}_\infty = cr_\infty$, one has the following two equivalent inequalities:

(i)
$$r_\infty < \frac{2}{\sqrt{a_j(f(0))}|f_j'(0)|}$$

(ii)
$$\hat{r}_\infty < \frac{2}{\sqrt{a_j(f(0))}|\frac{df_j}{d\hat{z}}(0)|}.$$

Therefore, by choosing c in such a way that $\pi\xi(0) = a_j(f(0))|f_j'(0)|^2 = 1$, the requirement reduces to $r_\infty < 2$.

In this situation one has

$$M(0) = \frac{1}{4\pi} \log \xi(0) = -\frac{1}{4\pi \log \pi},$$

so that by (3.2),

$$0 \leq Q(r) \leq \frac{1}{4\pi} \log \Phi(r) + \frac{1}{4\pi} \log \pi.$$

Hence, by letting $\hat{Q}(r) = 2\pi Q(r)$ one has

(3.3)
$$0 \leq \hat{Q}(r) \leq \frac{1}{2} \log \{\pi\Phi(r)\}.$$

We observe the following:

(3.4)
$$\begin{cases} \pi\Phi(r) \geq 1 \\ B(r) \geq 4\pi \int_0^r \frac{t}{\pi} dt = 2r^2 \\ \hat{Q}(r) = \int_0^r B(t)\frac{dt}{t} \geq r^2. \end{cases}$$

Lemma. $\pi\Phi(r) \geq r\hat{Q}(r)^4$ if $\pi \geq \sqrt{2}$.

Proof. Assume on the contrary that $\pi\Phi(r) < r\hat{Q}(r)^4$ holds for some $r \geq \sqrt{2}$. Then (3.3) implies

$$\hat{Q}(r) < 2\log\hat{Q}(r) + \frac{1}{2}\log r$$

so that

$$\hat{Q}(r) - 2\log\hat{Q}(r) < \frac{1}{2}\log r.$$

Since $r \geq \sqrt{2}$, (3.4) implies that $\hat{Q}(r) - \log\hat{Q}$ is nondecreasing in this range of r because

$$\frac{d}{d\hat{Q}}(\hat{Q} - 2\log\hat{Q}) = 1 - \frac{2}{\hat{Q}} \geq 0.$$

Hence

$$r^2 - 2\log r^2 \leq \hat{Q}(r) - 2\log\hat{Q}(r) < \frac{1}{2\log r}$$

by (3.4), so that we obtain

(3.5) $$r^2 - \frac{9}{2}\log r < 0.$$

On the other hand, since

$$\frac{d}{dt}\left(t^2 - \frac{9}{2}\log t\right) = \frac{2}{t}\left(t^2 - \frac{9}{4}\right),$$

the function $t^2 - \frac{9}{2}\log t$ is nondecreasing on the interval $[\frac{3}{2}, \infty)$ and takes its minimum on $[\sqrt{2}, \infty)$ at $t = \frac{3}{2}$.

As a result we obtain

$$r^2 - \frac{9}{2}\log r \geq \left(\frac{3}{2}\right)^2 - \frac{9}{2}\log\frac{3}{2} = \frac{9}{2}\left(\frac{1}{2} - \log\left(1 + \frac{1}{2}\right)\right) > 0,$$

which contradicts (3.5).

Lemma. *If $r \geq \sqrt{2}$,*

(i) $\pi r\Phi(r) \geq B(r)^2$

or

(ii) $B(r) \geq r\hat{Q}(r)^2.$

Proof. Suppose that $\pi r\Phi(r) < B(r)^2$ and $B(r) < r\hat{Q}(r)^2$ hold at the same time for some $r \geq \sqrt{2}$. Then

$$\pi r\Phi(r) < (r\hat{Q}(r)^2)^2 = r^2\hat{Q}(r)^4$$

would hold for such r, which means that $\pi\Phi(r) < r\hat{Q}(r)^4$, contradicting the preceding lemma.

Consequently, for $r \geq \sqrt{2}$, either

$$dr = \frac{dB(r)}{4\pi r \Phi(r)} \leq \frac{dB(r)}{4B(r)^2} \quad \text{(case i)}$$

or

$$dr = \frac{r\,d\hat{Q}(r)}{B(r)} \leq \frac{d\hat{Q}(r)}{\hat{Q}(r)^2} \quad \text{(case ii)}$$

holds. Here we have used the equalities $dB(r) = 4\pi\Phi(r)r\,dr$ and $d\hat{Q}(r) = B(r)\frac{dr}{r}$, which hold by definition. Accordingly,

$$dr \leq \frac{dB}{4B^2} + \frac{d\hat{Q}}{\hat{Q}^2}$$

holds if $r \geq \sqrt{2}$. Since $B(r)$ and $\hat{Q}(r)$ are monotonically increasing, this inequality leads to

$$r - \sqrt{2} = \int_{\sqrt{2}}^{r} dr \leq \frac{1}{4B(\sqrt{2})} + \frac{1}{\hat{Q}(\sqrt{2})} \leq \frac{1}{16} + \frac{1}{8},$$

$r \leq \sqrt{2} + \frac{1}{16} + \frac{1}{8} < 2$ and the desired conclusion $r_\infty < 2$.

4   Holomorphic Maps to the Riemann Sphere

We consider a holomorphic map $f : \Delta_{r_\infty} \to \mathbb{P}_1$. Since $\mathbb{P}_1 = \{w \in \mathbb{C}\} \cup \{\infty\}$, f is a meromorphic function on Δ_{r_∞}. We shall fix a canonical Kähler form

$$\omega = \frac{\sqrt{-1}}{2\pi} \frac{dw \wedge d\overline{w}}{(1 + |w|^2)^2}$$

on $\mathbb{P}_1$ (cf. Sect. 1). With respect to the stereographic projection from the sphere of radius $\frac{1}{2}$ to the w-plane, we shall denote the points on the sphere corresponding to $w, a \in \mathbb{C}$ by $\hat{w}, \hat{a}$. Then the length of the line segment connecting $\hat{w}$ and $\hat{a}$ is written as

$$\frac{|w - a|}{\sqrt{(1 + |w|^2)(1 + |a|^2)}}.$$

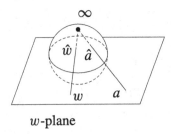

$$w\text{-plane}$$

Definition. $u = a(z) = \frac{1}{4\pi} \log \frac{(1+|f(z)|^2)(1+|a|^2)}{|f(z)-a|^2}$.

This function is $\frac{1}{2\pi}$ log of the reciprocal of the euclidean distance between $\widehat{f(z)}$ and $\hat{a}$. Since the distance is invariant under the rotation, $u_a(z)$ is naturally extended as a C^∞ function on $\Delta_{r_\infty} \times \mathbb{P}_1$, where the local expression on a neighborhood of ∞ is derived from the transformation $w \to 1/w$.

Note that

$$(4.1) \qquad\qquad dd^\perp u_a(z) = f^*\omega$$

holds, for the left hand side $= 2\sqrt{-1}\partial\bar{\partial}u_a(z) = \frac{\sqrt{-1}}{2\pi}\partial\bar{\partial}\log(1+|f(z)|^2) = \frac{\sqrt{-1}}{2\pi}\frac{|f'(z)|^2}{(1+|f(z)|^2)^2} dz \wedge d\bar{z} = f^*\omega$.

Notation

$$A(r) = \int_{\Delta_r} f^*\omega,$$

$$T(r) = \int_0^r A(t)\frac{dt}{2\pi t} \quad \text{(the order function)},$$

$$m(r,a) = \frac{1}{2\pi}\int_0^{2\pi} u_a(re^{i\theta})\,d\theta \quad \text{(the average over } \partial\Delta_r).$$

Lemma. (4.2) $m(r,a)$ *is continuous in* (r,a) *and piecewise smooth* (C^∞) *with respect to* r.

Proof. Let $\zeta_1(a), \ldots, \zeta_h(a), \ldots$ be the roots of $f(z) - a = 0$, repeated with multiplicity. Similarly, let $\gamma_h = \zeta_h(\infty)$, $h = 1, 2, \cdots$ be the poles of $f(z)$. It suffices to verify the assertion for $|a - a_0| < \varepsilon$ and $0 \le r < R$ for any $a_0 \in \mathbb{C}$ and R satisfying $R \neq |\zeta_h(a_0)|$, $h = 1, 2, \cdots$. In this setting,

$$f(z) - a = \frac{\prod_{|\zeta_h|<R}(z - \zeta_h(a))}{\prod_{|\gamma_h|<R}(z - \gamma_h)}g(z,a),$$

where $g(z,a)$ is holomorphic in (z,a) for $z \in \Delta_R$ and $|a - a_0| < \varepsilon$ for sufficiently small ε, and $g(z,a) \neq 0$ in this range. Therefore

$$u_a(z) = \frac{1}{4\pi} \log \frac{(\prod |z - \gamma_h| + \prod |z - \zeta_h(a)|^2 \cdot |g|^2)(1 + |a|^2)}{\prod |z - \zeta_h(a)|^2 \cdot |g|^2}$$

$$= -\frac{1}{4\pi} \sum_{|\zeta_h(a)| < R} \log |z - \zeta_h(a)|^2 + \{C^\infty \ function\}.$$

Since $\zeta_h(a)$ are continuous for sufficiently small ε, $m(r, a)$ is continuous in (r, a) and piecewise smooth in r.

Definition. $n(r, a) =$ the number of the roots of $f(z) - a = 0$ in Δ_r counted with multiplicity.

Definition. $N(r, a) = \int_0^r n(t, a) \frac{dt}{2\pi t}$ (the counting function) if $f(0) \neq a$.

Since

$$A(r) = \int_{\Delta_r} f^*\omega = \int_{\Delta_r} dd^\perp u_a$$

$$= \int_{\partial \Delta_r} d^\perp u_a - \sum_{\zeta_h(a) \in \Delta_r} \oint_{\zeta_h(a)} d^\perp u_a$$

$$= \int_0^{2\pi} \frac{\partial}{\partial r} u_a(re^{i\theta} r \, d\theta) + n(r, a)$$

$$\left(\oint_{\zeta_h(a)} d^\perp u_a = \oint \frac{\partial u_a}{\partial \nu} \, ds = 1 \right)$$

$$= 2\pi r \frac{d}{dr} m(r, a) + n(r, a),$$

we obtain the following.

Theorem. (4.3) **(The first main theorem)** If $f(0) \neq 0$,

$$N(r, a) + m(r, a) = T(r) + m(0, a).$$

(4.4) *Remark 1.* Since $A(t)$ is increasing in t,

$$T(r) = \int_0^r A(t) \frac{dt}{2\pi t} \geq A(r_0) \int_{r_0}^r \frac{dt}{2\pi t} \geq \frac{A(r_0)}{2\pi} (\log r - \log r_0),$$

so that

$$\liminf_{r \to \infty} \frac{T(r)}{\log r} > 0.$$

(4.5) *Remark 2.* $T(r) = \int_0^r A(t) \frac{dt}{2\pi t} = \int_{w \in \mathbb{P}_1} \{\int_0^r n(t, w) \frac{dt}{2\pi t}\} \omega(w) = \int_{w \in \mathbb{P}_1} N(r, w) \omega(w)$.

Example. If $f(z)$ is a rational function of degree n, $n(t, a) = n$ holds for $t > r_0$ if r_0 is sufficiently large. Hence $N(r, a) \sim \frac{n}{2\pi} \log r$ and $T(r) \sim \frac{n}{2\pi} \log r$.

Remark. Conversely, $f(z)$ is a rational function if $T(r) = O(\log r)$.

Remark 3. If $f(0) = a$, then the theorem holds for the following correction of $N(r, a)$:

$$N(r, a) = \int_0^r \{n(t, a) - n(0, a)\} \frac{dt}{2\pi t} + \frac{n(0, a)}{2\pi} \log r$$

or

$$N(r, a) = \int_{r_0}^r n(t, a) \frac{dt}{2\pi t} \qquad (r_0 > 0).$$

5   Defect Relation

We consider a holomorphic map $f : \mathbb{C} \to \mathbb{P}_1$ by letting $r_\infty = \infty$.

Definition. (5.1) *The quantity*

$$\delta(a) = \liminf_{r \to +\infty} \frac{m(r, a)}{T(r)}$$

*is called the **defect** of f at a.*

Definition. *An exceptional value of f if $f(z) \neq a$ holds for all $z \in \mathbb{C}$.*

Remark. By the first main theorem (4.3),

$$\frac{m(r, a)}{T(r)} = 1 - \frac{N(r, a)}{T(r)} + \frac{m(0, a)}{T(r)}.$$

Hence

$$\delta(a) = 1 - \limsup_{r \to +\infty} \frac{N(r, a)}{T(r)},$$

so that $\delta(a) = 1$ holds if a is an exceptional value of f. Note also that $0 \leq \delta \leq 1$.

Definition. $\delta_1 = \liminf_{r \to +\infty} \frac{N_1(r)}{T(r)}$.

Given $a_1, \ldots, a_q \in \mathbb{P}_1$, the following relation holds between $\delta(a_1), \ldots, \delta(a_q)$ and δ_1.

Theorem. (5.2) **(Defect relation)** *Let $f : \mathbb{C} \to \mathbb{P}_1$ be a nonconstant holomorphic map satisfying $f'(0) \neq 0$, $f(0) = a_k, k = 1, \ldots, q$. Then*

$$\sum_{k=1}^{q} \delta(a_k) + \delta_1 \leq 2.$$

Corollary 1. (Picard) *If f is nonconstant, then f admits at most two exceptional values.*

Proof. If $a_1, \ldots, a_q$ are exceptional values of f, then $\delta(a_k) = 1$ for all k. Hence $q \leq 2$ by the theorem.

Corollary 2. $\sum_{a \in \mathbb{P}_1} \delta(a) + \delta_1 \leq 2.$

Proof of Theorem (Nevanlinna [18], Wu [25]).

Definition.

$$\rho(w) = \frac{1}{d_\lambda} \left(\prod_{k=1}^{q} \frac{(1 + |w|^2)(1 + |a_k|^2)}{|w - a_k|} \right)^\lambda.$$

Here $0 < \lambda < 1$ and α_λ is a positive number satisfying

$$\int_{w \in \mathbb{P}_1} \rho(w)\omega(w) = 1.$$

Then $\rho(w)$ is C^∞ on $\mathbb{P}_1 \setminus \{a_1, \ldots, a_q\}$ and $\rho(a_k) = +\infty$. On a neighborhood of a_k,

$$\rho(w) = \frac{1}{t^{2\lambda}} \times \{C^\infty \text{ function}\},$$

where $t = |w - a_k|$, so that

$$\int_{|w - a_k| < \varepsilon} \rho(w)\omega(w) \sim \frac{1}{\pi} \int_0^{2\pi} \int_0^\varepsilon \frac{t\, dt\, d\theta}{t^{2\lambda}}$$

$$= 2 \int_0^\varepsilon t^{1-2\lambda} dt \sim \varepsilon^{2-2\lambda}.$$

Hence $\rho(w)\omega$ is integrable on $\mathbb{P}_1$.

Definition. $\Psi(r) = \int_{\Delta_r} f^*(\rho\omega).$

$$\Psi(r) = \int_{\Delta_r} \rho(f(z)) f^* \omega = \int_{\Delta_r} \rho(f(z)) \xi(z) \frac{\sqrt{-1}}{2} dz \wedge d\bar{z}$$

$$= \int_{\Delta_r} \rho(f(z)) \xi(z) \frac{\sqrt{-1}}{2} dz \wedge d\bar{z} = \int_{\Delta_r} \rho(f(te^{i\theta})) \xi(te^{i\theta}) t\, dt\, d\theta$$

$$= \int_0^r t\, dt \int_0^{2\pi} \rho(f(te^{i\theta})) \xi(te^{i\theta})\, d\theta.$$

Here we put

$$\xi(z) = \frac{|f'(z)|^2}{\pi(1 + |f(z)|^2)^2}. \qquad \text{(cf.Sect.2).}$$

Definition. $\Phi(t) = \frac{1}{2\pi} \int_0^{2\pi} \rho(f(te^{i\theta}))\xi(te^{i\theta})\, d\theta.$

By the preceding computation,

$$\Psi(r) = \int_0^r 2\pi t \Phi(t)\, dt.$$

Lemma. $\lambda \sum_{k=1}^q m(r, a_k) + M(r) \leq \frac{1}{4\pi} \log \alpha_\lambda + \frac{1}{4\pi} \log \Phi(r).$

Proof. By the definition of ρ,

$$\log \rho(f(z)) = -\log \alpha_\lambda + 4\pi\lambda \sum_{k=1}^q u_{a_k}(z).$$

Averaging on $\partial\Delta_r$ yields

$$\frac{1}{2\pi} \int_0^{2\pi} \log \rho(f(re^{i\theta}))\, d\theta = -\log \alpha_\lambda + 4\pi\lambda \sum_{k=1}^q m(r, a_k).$$

Hence

$$\lambda \sum_{k=1}^q m(r, a_k) + M(r)$$

$$= \frac{1}{4\pi} \log \alpha_\lambda + \frac{1}{2\pi} \int_0^{2\pi} \frac{1}{4\pi} \log\left(\rho(f(z))\xi(z)\right) d\theta$$

$$\leq \frac{1}{4\pi} \log \alpha_\lambda + \frac{1}{4\pi} \log \left\{ \frac{1}{2\pi} \int_0^{2\pi} \rho(f)\xi\, d\theta \right\}$$

$$= \frac{1}{4\pi} \log \alpha_\lambda + \frac{1}{4\pi} \log \Phi(r).$$

Combining the lemma with Theorem (2.4) and applying the equality

$$\int_0^t B(t) \frac{dt}{2\pi t} = -2T(r) \qquad (R = 4)$$

one has

$$\lambda \sum_{k=1}^q m(r, a_k) + N_1(r) \leq 2T(r) + \frac{1}{4\pi} \log \Phi(r) + C_\lambda,$$

where C_λ is a constant depending only on λ. Dividing both sides of the inequality by $T(r)$ and taking the lower limit, we obtain

$$\lambda \sum_{k=1}^q \delta(a_k) + \delta_1 \leq 2 + \frac{1}{4\pi} \liminf_{r \to +\infty} \frac{\log \Phi(r)}{T(r)}.$$

Therefore, by letting $\lambda \to 1$, the proof of Theorem (5.2) is reduced to showing

(5.4) $$\liminf_{r \to +\infty} \frac{\log \Phi(r)}{T(r)} = 0.$$

Notation

$$Q(r) = \int_0^r \Psi(t) \frac{dt}{2\pi t}.$$

Since

$$\Psi(t) = \int_{\Delta_t} f^*(\rho\omega) = \int_{a \in \mathbb{P}_1} n(t, a)\rho(a)\omega(a),$$

one has

$$Q(r) = \int_{a \in \mathbb{P}_1} N(r, a)\rho(a)\omega(a).$$

By the first main theorem (4.3),

$$N(r, a) \le T(r) + m(0, a)$$

since $m(r, a) \ge 0$.

   Hence

$$\int_{a \in \mathbb{P}_1} N_1(r, a)\rho(a)\omega(a) \le T(r) + m_0,$$

where m_0 is a constant which does not depend on λ. (In fact, since $m(0, a)$ is continuous in a, by Lemma (4.2) there exists a constant m_0 satisfying $m(0, a) \le m_0$.)

   Thus we obtain

(5.5) $$Q(r) \le T(r) + m_0.$$

Lemma.

$$\liminf_{r \to +\infty} \frac{\Phi(r)}{Q(r)^6} = 0.$$

Proof. If

$$\liminf_{r \to +\infty} \frac{\Phi(r)}{Q(r)^6} > 0,$$

there would exist $\varepsilon > 0$ and $r_1 > 0$ such that the following holds for all $r > r_1$:

(5.6) $$\frac{\Phi(r)}{Q(r)^6} \ge \varepsilon^4 > 0.$$

From this it follows that either

(i) $r^{3/2}\Phi(r) \geq \varepsilon\Psi(r)^3$

   or

(ii) $\Psi(r) \geq \varepsilon r^{1/2}Q(r)^2$

holds if $r > r_1$.

In fact, if neither of (i) or (ii) is true, $\Phi(r) < \varepsilon^4 Q(r)^6$ would hold. But it contradicts (5.6).

We put

$$\hat{\Psi}(t) = \frac{1}{2\pi}\Psi(t).$$

Then $d\hat{\Psi} = \Phi(t)t\,dt$ and $dQ = \hat{\Psi}(t)\frac{dt}{t}$ by definition.

Therefore

$$2d\sqrt{r} = \frac{dr}{\sqrt{r}} = \frac{d\hat{\Psi}}{r^{3/2}\Phi} \leq \frac{d\hat{\Psi}(r)}{\varepsilon\hat{\Psi}(r)^3}$$

if (i) is true, and

$$2d\sqrt{r} = \frac{dQ}{r^{1/2}\hat{\Psi}} \leq \frac{dQ(r)}{\varepsilon Q(r)^2}$$

if (ii) holds.

Since $\hat{\Psi}$ and Q are increasing in r,

$$2d\sqrt{r} \leq \frac{d\hat{\Psi}(r)}{\varepsilon\hat{\Psi}(r)^3} + \frac{dQ(r)}{\varepsilon Q(r)^2}$$

holds if $r > r_1$.

Hence

$$2(\sqrt{r} - \sqrt{r_1}) = \int_{r_1}^{r} 2d\sqrt{t} \leq \frac{1}{\varepsilon}\left(\frac{1}{2\hat{\Psi}(r_1)^2} + \frac{1}{Q(r_1)}\right) < +\infty,$$

so that $r < const.$ which is an absurdity.

Proof of Theorem 5.2. By the above lemma, one can find $r_2 > 0$ such that $\Phi(r) \leq Q(r)^6$ holds for all $r > r_2$.

Hence, from the inequality (5.5) it follows that

$$\liminf_{r\to+\infty} \frac{\log \Phi(r)}{T(r)} \leq \liminf_{r\to+\infty} \frac{6\log Q(r)}{T(r)}$$

$$\leq 6\liminf_{r\to+\infty} \frac{\log(T(r) + M_0)}{T(r)} = 0.$$

(Note that $T(r) \to +\infty$ as $r \to +\infty$ by Remark (4.4).)

Chapter 2
Schwarz–Kobayashi Lemma

Abstract We shall extend the classical Schwarz lemma to the case of several variables.

Keywords Schwarz–Kobayashi lemma · Canonical bundle · Algebraic manifolds of general type · Hyperbolic measure

We shall extend the classical Schwarz lemma to the case of several variables.

1 The Schwarz–Kobayashi Lemma

$\mathbb{D}$ will stand for the unit disc = $\{z \in \mathbb{C} | |z| < 1\}$.

(1.1) **The Schwarz lemma.** For any holomorphic map $f : \mathbb{D} \to \mathbb{D}$ satisfying $f(0) = 0$,

(i) $|f(z)| \leq |z|, \quad z \in \mathbb{D}$

and

(ii) $|f'(0)| \leq 1$

hold.

Utilizing the Kähler form $\omega = \frac{\sqrt{-1}}{2} \frac{dw \wedge d\overline{w}}{(1-|w|^2)^2}$, the lemma can be stated as

$$f^* \omega_W \leq \omega_Z.$$

© The Author(s) 2017
K. Kodaira, *Nevanlinna Theory*, SpringerBriefs in Mathematics,
https://doi.org/10.1007/978-981-10-6787-7_2

The Schwarz lemma will be extended to n-variables in this form.

Notation

$\mathbb{C}^n$: n-dimensional complex vextor space $=\{(z_1, \ldots, z_n)\}$,
Δ_r : n-dimensional disc $= \{z \in \mathbb{C}^n \mid |z| < r\}$,
$\partial \Delta_r$: the boundary of $\Delta_r = \{z \in \mathbb{C}^n \mid |z| = 1\}$,

where $|z| = \sqrt{\sum_{\alpha=1}^{n} |z_\alpha|^2}$,

$dV(z)$: the volume form of $\mathbb{C}^n = (\frac{\sqrt{-1}}{2})^n \, dz_1 \wedge d\bar{z}_1 \wedge \cdots \wedge dz_n \wedge d\bar{z}_n$.

(1.2) Definition.

$$v_r = \frac{r^2}{(r^2 - |z|^2)^{n+1}} \, dV(z)$$

$$\mu_r(z) = \frac{r^2}{(r^2 - |z|^2)^{n+1}}$$

$$\omega_r(z) = \frac{\sqrt{-1}}{n+1} \partial \bar{\partial} \log \mu_r(z).$$

Proposition. v_r *is invariant under the biholomorphic maps from Δ_r to Δ_r.*

Proof. Clearly it suffices to prove it for $r = 1$.

(i) Let $g : \Delta_1 \to \Delta_1$ be a biholomorphic map with $g(0) = 0$. Then g is a unitary transformation (cf. Bochner–Martin [2]). Since $|z|$ and $dV(z)$ are unitary invariant, so is v_1.

(ii) Given $b = (b_1, 0, \ldots, 0) \in \Delta_1)$, let $g_b : z \to z' = g_b(z)$ be the biholomorphic transformation of Δ_1 defined by

$$\begin{cases} z_1' = \dfrac{z_1 - b_1}{1 - (\bar{b}z)} \\ z_\alpha' = \dfrac{\sqrt{1 - |b|^2}}{1 - (\bar{b}z)} z_\alpha, \quad \alpha = 2, \ldots, n. \end{cases}$$

Here we put $(\bar{b}z) = \bar{b}_1 z_1$. In particular $g_b(b) = 0$.
Then

$$1 - |z'|^2 = \frac{(1 - |b|^2)(1 - |z|^2)}{|1 - (\bar{b}z)|^2},$$

$$dz_1' = \frac{1 - |b|^2}{(1 - (\bar{b}z))^2} dz_1$$

and

$$dz'_\alpha = \frac{\sqrt{1 - |b|^2}}{1 - (\bar{b}z)} dz_\alpha + \frac{(\sqrt{1 - |b|^2})\bar{b}_1 z_\alpha}{(1 - (\bar{b}z))^2} dz_1$$

for $2 \le \alpha \le n$.

Hence

$$g_b^* v_1 = \frac{dV(z')}{(1 - |z'|^2)^{n+1}}$$

$$= \frac{(1 - |b|^2)^{n+1}}{|1 - (\bar{b}z)|^{2(n+1)}} \left[\frac{|1 - (\bar{b}z)|^2}{(1 - |b|^2)(1 - |z|^2)} \right]^{n+1} dV(z)$$

$$= \frac{dV(z)}{(1 - |z|^2)^{n+1}} = v_1.$$

(iii) For any biholomorphic map $g : \Delta_1 \to \Delta_1$, let u_2 be a unitary transformation which maps $g^{-1}(0)$ to $b = (b_1, 0, \ldots, 0)$ with $b_1 = |g^{-1}(0)|$. Then $g_b \circ u_2 \circ g^{-1}$ is a unitary transformation since $g_b \circ u_2 \circ g^{-1}(0) = 0$. Letting $u_1^{-1} = g_b \circ u_2 \circ g^{-1}$, one has $g = u_1 \circ g_b \circ u_2$. Therefore, in view of (i) and (ii), v_1 is invariant under g.

Some properties of ω_r defined by

$$\omega_r = -\partial\bar{\partial} \log \left(r^2 - \sum_{\alpha=1}^{n} z_\alpha \bar{z}_\alpha \right)$$

will be described below.

We note that

$$\omega_r = \sqrt{-1} \left\{ \frac{\sum_{\alpha=1}^{n} dz_\alpha \wedge d\bar{z}_\alpha}{r^2 - |z|^2} + \frac{(\sum_{\alpha=1}^{n} \bar{z}_\alpha dz_\alpha) \wedge (\sum_{\beta=1}^{n} z_\beta d\bar{z}_\beta)}{(r^2 - |z|^2)^2} \right\}$$

$$= \sqrt{-1} \sum_{\alpha,\beta} g_{\alpha\beta} dz_\alpha \wedge d\bar{z}_\beta,$$

where

$$g_{\alpha\beta} = \frac{(r^2 - |z|^2)\delta_{\alpha\beta} + \bar{z}_\alpha z_\beta}{(r^2 - |z|^2)^2},$$

$$\delta_{\alpha\beta} = \begin{cases} 1 & (\alpha = \beta) \\ 0 & (\alpha \neq \beta). \end{cases}$$

Clearly

$$\sum_{\alpha,\beta} g_{\alpha\beta} \xi_\alpha \bar{\xi}_\beta = \frac{(r^2 - |z|^2)|\xi|^2 + |(z\bar{\xi})|^2}{(r^2 - |z|^2)^2}$$

holds for any $(\xi_1, \ldots, \xi_n) \in \mathbb{C}^n \setminus \{0\}$. Hence $\omega_r > 0$ (cf. (1.5)), so that ω_r is a Kähler form on Δ_r.

(1.3) **Lemma.** $\omega_r^n = 2^n n! v_r$.

Proof. According to the above notaion one has

$$\omega_r^n = (\sqrt{-1})^n \sum g_{\alpha_1 \beta_1} \cdots g_{\alpha_n \beta_n} \, dz_{\alpha_1} \wedge d\bar{z}_{\beta_1} \wedge \cdots \wedge dz_{\alpha_n} \wedge d\bar{z}_{\beta_n}$$
$$= \det(g_{\alpha\beta}) \cdot 2^n n! \, dV(z).$$

Letting $\sigma = r^2 - |z|^2$ and $D = \det(g_{\alpha\beta})$, one has

$$\sigma^{2n} D = \begin{vmatrix} \sigma + z_1\bar{z}_1 & z_2\bar{z}_1 & \cdots & z_n\bar{z}_1 \\ \bar{z}_2 z_1 & \sigma + z_2\bar{z}_2 & \cdots \cdot \\ \cdots & \cdots & \cdots \\ \bar{z}_n z_1 & \cdots \cdots & \sigma + z_n\bar{z}_n \end{vmatrix}$$

so that

$$\bar{z}_1 \sigma^{2n} D = \begin{vmatrix} (\sigma + |z_1|^2)\bar{z}_1 & z_2\bar{z}_1 & \cdots & z_n\bar{z}_1 \\ |z_1|^2\bar{z}_2 & \sigma + z_2\bar{z}_2 & \cdots \cdot \\ \cdots & \cdots & \cdots \\ |z_1|^2\bar{z}_n & \cdots \cdots & \sigma + z_n\bar{z}_n \end{vmatrix}.$$

By adding (the second column) $\times \bar{z}_2 + \cdots +$ (the n-th column)$\times \bar{z}_n$ to the first column, and noting that $(\sigma + |z|^2)\bar{z}_\alpha = r^2\bar{z}_\alpha$, we obtain

$$\bar{z}_1 \sigma^{2n} D = \begin{vmatrix} r^2\bar{z}_1 & z_2\bar{z}_1 & \cdots & z_n\bar{z}_1 \\ r^2\bar{z}_2 & \sigma + z_2\bar{z}_2 & \cdots \cdot \\ \cdots & \cdots & \cdots \\ r^2\bar{z}_n & \cdots \cdots & \sigma + z_n\bar{z}_n \end{vmatrix}$$

$$= r^2 \begin{vmatrix} \bar{z}_1 & & & \\ & \sigma & & \\ & & \ddots & \\ & & & \sigma \end{vmatrix} = r^2\sigma^{n-1}\bar{z}_1.$$

Namely $\sigma^{2n} D = r^2\sigma^{n-1}$.

Therefore

$$D = \frac{r^2}{\sigma^{n+1}} - \mu_r(z)$$

which means $\omega_r^n = 2^n n! v_r$.

Let $f : \Delta_r \to \Delta_1$ be any holomorphic map.

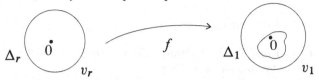

Theorem. (1.4) **(The Schwartz–Kobayashi lemma; cf. [11])**

$$f^* v_1 \leq v_r.$$

Proof. Let $f : z \to (f_1(z), \ldots, f_n(z))$ and $J(z) = \det\left(\frac{\partial f_\alpha(z)}{\partial z_\beta}\right)$. Then

$$f^* v_1 = \mu_1(f(z)) \, dV(f(z)) = \mu_1(f(z)) |J(z)|^2 \, dV(z).$$

We define a C^∞ function $\xi(z)$ on Δ_r by $\xi(z) = \mu_1(f(z)) |J(z)|^2$.

Let $0 < t < r$. Then the function

$$\frac{\xi(z)}{\mu_t(z)} = \frac{(t^2 - |z|^2)^{n+1}}{t^2} \xi(z)$$

is C^∞ on $\Delta_t \cup \partial \Delta_t$, nonnegative there, and takes the value 0 on $\partial \Delta_t$. Hence $\frac{\xi(z)}{\mu_t(z)}$ takes its maximum at some point, say $z_0 \in \Delta_t$.

Accordingly, letting $\ell(z) = \log \xi(z) - \log \mu_t(z)$ one has $\ell(z) \leq \ell(z_0)$ for any $z \in \Delta_t$. Therefore, by differentiation

$$\partial \bar{\partial} \ell(z) = \sum_{\alpha, \beta} \ell_{\alpha\beta} \, dz_\alpha \wedge d\bar{z}_\beta,$$

the matrix $(\ell_{\alpha\beta}(z_0))$ is known to be negative semidefinite.

Definition. (1.5) *A $(1, 1)$ form $\omega = \sqrt{-1} \sum_{\alpha, \beta} h_{\alpha\beta}(z) \, dz_\alpha \wedge d\bar{z}_\beta$ is said to be positive definite (resp. positive semidefinite) at z, written as $\omega > 0$ (resp. $\omega \geq 0$), if the matrix $(h_{\alpha\beta}(z))$ is positive definite (resp. positive semidefinite). The negativity of $(1, 1)$ forms are similarly defined.*

In this terminology, the above-mentioned property of ℓ can be written as

$$[\sqrt{-1}\partial\bar{\partial}\ell]_{z=z_0} \leq 0.$$

Further, since

$$\sqrt{-1}\partial\bar{\partial} \log \xi(z) = \sqrt{-1}\partial\bar{\partial} \log \left(\mu_1(f(z))|J(z)|^2\right)$$
$$= \sqrt{-1}\partial\bar{\partial} \log \mu_1(f(z)) = f^* \omega$$

and $\omega_t = \sqrt{-1}\partial\bar{\partial} \log \mu_t$, one has $[f^*\omega_1 - \omega_t]_{z=z_0} \leq 0$. Hence

(1.6) $$(f^*\omega_1)^n_{z=z_0} \leq (\omega_t)^n_{z=z_0}$$

holds by the following lemma.

Lemma. *Let $(h_{\alpha\beta})$ and $(g_{\alpha\beta})$ be positive definite Hermitian matrices such that $(h_{\alpha\beta} - g_{\alpha\beta})$ is negative semidefinite. Then*

$$\det(h_{\alpha\beta}) \leq \det(g_{\alpha\beta}).$$

Proof. Let us put

$$H = \left\{ z \in \mathbb{C}^n \ \middle|\ \sum_{\alpha,\beta} h_{\alpha\beta} z_\alpha \overline{z_\beta} < 1 \right\}$$

and

$$G = \left\{ z \in \mathbb{C}^n \ \middle|\ \sum_{\alpha,\beta} g_{\alpha\beta} z_\alpha \overline{z_\beta} < 1 \right\}.$$

Then $G \subset H$ by the assumption. Thus evaluations of the volume as

$$\int_G dV(z) = \frac{1}{\det(g_{\alpha\beta})} \frac{\pi^n}{n!}$$

and

$$\int_H dV(z) = \frac{1}{\det(h_{\alpha\beta})} \frac{\pi^n}{n!}$$

yield $\det(h_{\alpha\beta}) \leq \det(g_{\alpha\beta})$.

Combining the inequality (1.6) with Lemma (1.3), one has

$$(f^*v_1)_{z=z_0} \leq (v_t)_z = z_0,$$

which means

$$\xi(z_0)\, dV \leq \mu_t(z_0)\, dV.$$

Hence $\xi(z_0) \leq \mu_t(z_0)$ so that

$$\frac{\xi(z)}{\mu_t(z)} \leq \frac{\xi(z_0)}{\mu_t(z_0)} \leq 1.$$

Thus we obtain $\xi(z) \leq \mu_t(z)$, if $z \in \Delta_t$ and $0 < t < r$, so that by letting t tend to r one has $\xi(z) \leq \mu_r(z)$.

Consequently,

$$f^*v_1 = \xi(z)\, dV(z) \leq \mu_r(z)\, dV(z) = v_r.$$

Remark. Since the proof of Theorem 1.4 is based only on Definition (1.2) and Lemma (1.3), the theorem still holds true for more general cases, symmetric bounded domains for instance.

Example 1. Let $\mathbb{D}_r = \{z \in \mathbb{C}^n \mid |z_i| < r\}$ (called a polydisc) and set

$$
\begin{cases}
\tilde{v}_r = \tilde{u}_r(z)\, dV(z) \\
\tilde{\mu}_r(z) = \displaystyle\prod_{\alpha=1}^{n} \frac{r^2}{(r^2 - |z_\alpha|^2)^2} \\
\tilde{\omega}_r = \dfrac{\sqrt{-1}}{2} \partial\bar{\partial} \log \tilde{\mu}_r.
\end{cases}
$$

Then $\tilde{\omega}_r^n = 2^n n! \tilde{v}_r$. Hence, for any holomorphic map $f : \mathbb{D}_r \to \mathbb{D}_1$ one has $f^* \tilde{v}_1 \le \tilde{v}_r$.

Example 2. For any holomorphic map $f : \mathbb{D}_r \to \Delta_1$ one has

$$
f^* v_1 \le \left(\frac{2}{n+1} \right)^n \tilde{v}_r.
$$

2 Holomorphic Maps to Algebraic Manifolds of General Type

Let W be a (connected) complex manifold of dimension n.

Definition. *A **line bundle** F over W is a complex manifold F equipped with a surjective holomorphic $\varpi : F \to W$ satisfying the following conditions.*

(i) *There exists an open covering $W = \bigcup_j U_j$ such that $\varpi^{-1}(U_j) = U_j \times \mathbb{C}$ and $\varpi : U_j \times \mathbb{C} \to U_j$ is the projection to the first factor.*

(ii) *If $w \in U_j \cap U_k$, $(w, \zeta_j) \in U_j \times \mathbb{C}$ and $(w, \zeta_k) \in U_k \times \mathbb{C}$ represent the same point if and only if $\zeta_j = f_{jk}(w)\zeta_k$, where f_{jk} is a holomorphic function on $U_j \cap U_k$ such that $f_{jk}(w) \ne 0$.*

$f_{ik}(w) = f_{ij}(w)f_{jk}(w)$ if $w \in U_i \cap U_j \cap U_k$. The system of functions f_{jk} satisfying this relation is called a **1-cocycle**.

Definition. *A **holomorphic section** φ of F over W is a holomorphic map $\varphi : W \to F$ satisfying $\varpi\varphi(w) = w$.*

A holomorphic section φ is represented by $\varphi : w \to \varphi(w) = (w, \varphi_j(w))$ on U_j, where $\varphi_j(w) = f_{jk}(w)\varphi_k(w)$ holds if $w \in U_j \cap U_k$.

The complex vector space of holomorphhic sections of F will be denoted by $H^0(W, \mathcal{O}(F))$.

The cocyle $\{f_{jk}\}$ determines the line bundle F. By $F^m = F \otimes \cdots \otimes F$ we denote the line bundle determined by $\{f_{jk}^m\}$.

Proposition. *(2.1) Let W be a compact complex manifold of dimension n and let F be a line bundle over W. Then*

$$\dim H^0(W, \mathcal{O}(F^m)) \leq O(m^n).$$

Proof. (Siegel [20]) Since W is compact, there exists a finite open cover $W = \bigcup_{j=1}^{k} U_j$ such that

$$U_j = \{w_j = (w_j^1, \ldots, w_j^n) \in \mathbb{C}^n \mid |w_j^\alpha| < 1, \alpha = 1, \ldots, n\}.$$

We put

$$U_j^r = \{w_j \in U_j \mid |w_j^\alpha| < r, \alpha = 1, \ldots, n\} \qquad \text{for } 0 < r < 1.$$

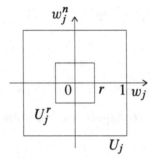

A holomorphic section $\varphi \in H^0(W, \mathcal{O}(F^m))$ will be identified with a system of holomorphic functions $\{\varphi_j(w); j = 1, \ldots, h\}$ satisfying $\varphi_j(w) = \{f_{jk}(w)\}^m \varphi_k(w)$ on $U_j \cap U_k$.

Notation

$$\|\varphi\|_r = \max_j \sup_{w \in U_j^r} |\varphi_j(w)|.$$

Expanding $\varphi_j(w)$ into a power series

$$\varphi_j(w) = \sum a_{j\nu_1 \ldots \nu_n} (w_j^1)^{\nu_1} \cdots (w_j^n)^{\nu_n}$$

we put

$$\varphi_j^\nu(w) = \sum_{\nu_1 + \cdots + \nu_n \leq \nu - 1} a_{j\nu_1 \ldots \nu_n} (w_j^1)^{\nu_1} \cdots (w_j^n)^{\nu_n}.$$

Definition. *Let $\psi^\nu = \{\psi_j^\nu; j = 1, \ldots, h\}$, where $\psi_j^\nu = \psi_j^\nu(w_j)$ are homogeneous polunomials of degree $\nu - 1$, and let $\mathcal{L}^\nu$ be the vector space consisting of such ψ^ν.*

Then, for any m and $\varphi \in H^0(W, \mathcal{O}(F^m))$, $\varphi^\nu = \{\varphi_j^\nu(w_j); j = 1, \ldots, h\} \in \mathcal{L}^\nu$ and the correspondence $\varphi \to \varphi^\nu$ is a homomorphism from $H^0(W, \mathcal{O}(F^m))$ to $\mathcal{L}^\nu$.

Lemma. *The above homomorphism $\varphi \to \varphi^\nu$ is injective for sufficiently large ν.*

Proof. We fix $0 < a < b < 1$ in such a way that $\bigcup_j U_j^a = W$. Then we choose a constant B such that $|f_{jk}(w)| \leq B < +\infty$ for $w \in U_j^b \cap U_k^b$ and for all j, k. Then, for any $w \in U_j^b$ one can find k such that $w \in U_k^a$, so that

$$|\varphi_j(w)| = |\{f_{jk}(w)\}^m \varphi_k(w)| \leq B^m \|\varphi\|_a$$

holds for any j.

Hence

$$(2.2) \qquad \|\varphi\|_b \leq B^m \|\varphi\|_a.$$

To prove the lemma, it suffices to show that $\varphi^\nu = 0$ implies $\varphi = 0$. Suppose that $\varphi^\nu = 0$. Then

$$\varphi_j(w_j) = \sum_{\nu_1 + \cdots + \nu_n \geq \nu} a_{j\nu_1 \ldots \nu_n} (w_j^1)^{\nu_1} \cdots (w_j^n)^{\nu_n}.$$

Hence $\varphi_j(tw_j) = t^\nu G_j(t, w_j)$ for some convergent power series G_j in (t, w_j).

For any $w_j \in U_j^a$, one can find $\epsilon > 0$ such that $tw_j \in U_j$ holds if $|t| < \frac{b}{a} + \epsilon$. Hence $G_j(t, w_j)$ is holomorphic in t for $|t| \leq \frac{b}{a}$. Let the maximum of $|G_j(t, w_j)|$ on $|t| = \frac{b}{a}$ be taken at t_0. Since $G_j(t, w_j)$ is holomorphic, $|G_j(t, w_j)| \leq |G_j(t_0, w_j)|$ for all $|t| \leq \frac{b}{a}$. In particular, for $t = 1 < \frac{b}{a}$ one has

$$|\varphi_j(w_j)| \leq \left| \frac{\varphi_j(t_0, w_j)}{t_0^\nu} \right|.$$

Since $|t_0| = \frac{b}{a}$, $t_0 w_j \in U_j^b$. Hence

$$|\varphi_j(w_j)| \leq \left(\frac{a}{b} \right)^\nu \|\varphi\|_b.$$

Combining this inequality with (2.1) one has

$$\|\varphi\|_a \leq \left(\frac{a}{b} \right)^\nu B^m \|\varphi\|_a.$$

Since $\left(\frac{a}{b} \right)^\nu B^m < 1$ for sufficiently large ν, we obtain $\|\varphi\|_a = 0$ or $\varphi = 0$. In other words, for any integer κ satisfying $\kappa > \frac{\log B}{\log \frac{b}{a}}$, the correspondence $\varphi \to \varphi^\nu$ is injective if $\nu \geq \kappa m$.

End of the proof of Proposition (2.1): By the lemma,

$$\dim H^0(W, \mathcal{O}(F^m))) \leq \dim \mathcal{L}^{\kappa m} \leq c\kappa^n m^n$$

for some constant c.

Let W be an n-dimensional projective algebraic manifold, i.e. an n-dimensional compact complex manifold embedded in some complex projective space $\mathbb{P}_N$. Let $W = \bigcup_j U_j$ be a covering by local coordinate neighborhoods and let $(w_j^1, \ldots, w_j^n)$ be the local coordinate on U_j.

Definition. (2.3) *The **canonical line bundle** of W, denoted by K, is the line bundle defined by the following 1-cocyle $\{J_{jk}\}$:*

$$J_{jk}(w) = \det \frac{\partial(w_k^1, \ldots, w_k^n)}{\partial(w_j^1, \ldots, w_j^n)}, \quad w \in U_j \cap U_k.$$

Definition. (2.4) *W is called an **algebraic manifold of general type** if*

$$\limsup_{m \to +\infty} \frac{\dim H^0(W, \mathcal{O}(K^m))}{m^n} > 0.$$

Let V be the hyperplane section of W, let $L = [V]$ be the line bundle associated to V, and let $L = \{l_{jk}\}$. Let $\{a_j\}$ be a metric of L (cf. §10), i.e. a_j is a C^∞ function on U_j satisfyng $a_j > 0$ and $a_j(w) = |l_{jk}(w)|^2 a_k(w)$ for $w \in U_j \cap U_k$. Note that $\partial\bar{\partial} \log a_j(w) = \partial\bar{\partial} \log a_k(w)$ for $w \in U_j \cap U_k$. Since L is the line bundle associated to the hyperplane section, it admits a metric $\{a_j\}$ satifying $\omega = \sqrt{-1}\partial\bar{\partial} \log a_j(w) > 0$.

$$W \subset \mathbb{P}_N$$

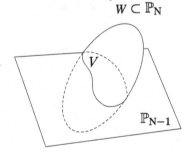

Lemma. *Let W be an algebraic manifold of general type. Then, for sufficiently large m,*

$$\dim H^0(W, \mathcal{O}(K^m \otimes L^{-1})) > 0.$$

Proof ([13], [15]). Let K_V be the restriction of K to V. Then

$$0 \longrightarrow \mathcal{O}(K^m \otimes L^{-1}) \longrightarrow \mathcal{O}(K^m) \longrightarrow \mathcal{O}(K_V^m) \longrightarrow 0 \quad \text{(exact sequence).}$$

From this sequence the following long exact sequence arises:

$$0 \longrightarrow H^0(W, \mathcal{O}(K^m \otimes L^{-1})) \longrightarrow H^0(W, \mathcal{O}(K^m)) \longrightarrow H^0(V, \mathcal{O}(K_V^m)) \longrightarrow \cdots.$$

By Proposition (2.1),

$$\dim H^0(V, \mathcal{O}(K_V^m)) = O(m^{n-1}).$$

Combining it with the assumption (cf. Definition (2.4)),

$$\limsup_{m \to +\infty} \frac{\dim H^0(W, \mathcal{O}(K^m \otimes L^{-1}))}{m^n} > 0.$$

Accordingly, $\dim H^0(W, \mathcal{O}(K^m \otimes L^{-1})) > 0$ for sufficiently large m.

Let us fix m so that $\dim H^0(W, \mathcal{O}(K^m \otimes L^{-1})) > 0$ and take $\varphi \in H^0(W, \mathcal{O}(K^m \otimes L^{-1}))$, $\varphi \neq 0$. In terms of the local expression $\varphi = \{\varphi_j(w)\}$,

$$\varphi_j(w) = J_{jk}(w)^m l_{jk}(w)^{-1} \varphi_k(w).$$

Hence

$$a_j(w)|\varphi_j|^2 = |J_{jk}|^{2m} a_k(w)|\varphi_k|^2,$$
$$(a_j|\varphi_j|^2)^{1/m} = |J_{jk}|^2 (a_k|\varphi_k^2|)^{1/m}.$$

Note that $dV(w_j) = |J_{jk}|^{-2} dV(w_k)$ with respect to the volume form

$$dV(w_j) = \left(\frac{\sqrt{-1}}{2}\right)^n dw_j^1 \wedge d\overline{w}_j^1 \wedge \cdots \wedge dw_j^n \wedge d\overline{w}_j^n.$$

Therefore

$$(a_j|\varphi_j|^2)^{1/m} dV(w_j) = (a_k|\varphi_k|^2)^{1/m} dV(w_k) \qquad \text{on } U_j \cap U_k.$$

Hence

(2.5) $$v = (a_j|\varphi_j|^2)^{1/m} dV(w_j)$$

is a well-defined volume form on W.

By letting

$$\omega = \sqrt{-1}\partial\bar{\partial} \log a_j = \sqrt{-1} \sum_{\alpha,\beta} g_{j\alpha\beta} \, dw_j^\alpha \wedge d\overline{w}_j^\beta$$

and $g_j(w) = \det(g_{j\alpha\beta}(w))$, we obtain

$$\omega^n = 2^n n! g_j(w) \, dV(w_j).$$

Since the ratio

$$\frac{2^n n! v}{\omega^n} = \frac{(a_j(w)|\varphi_j(w)|^2)^{1/m}}{g_j(w)}$$

is a continuous function on W, its maximum, say κ is taken somewhere. Then

(2.6) $$(a_j(w)|\varphi_j(w)|^2)^{1/m} \le \kappa g_j(w)$$

holds.

Theorem. (7.7) (**Schwarz–Kobayashi–Ochiai, cf. [13]**) *Let W be an algebraic manifold of general type and let $f : \Delta \to W$ be a holomorphic map whose Jacobian is not identically 0. Then*

$$f^* v \le (n+1)^n m^n \kappa v_r.$$

Here v is the volume form defined by (2.5) and

$$v_r = \frac{r^2 \, dV(z)}{(r^2 - |z|^2)^{n+1}}. \quad \text{(cf. (1.2))}$$

Proof. The proof goes similarly to Theorem (1.4). First, letting $f : z \to w_j = f_j(z)$ for $z \in f^{-1}(U_j)$, we put

$$J_j(z) = \det \frac{\partial(w_j^1, \ldots, w_j^n)}{\partial(z_1, \ldots, z_n)}.$$

Setting $f^* v = \xi(z) \, dV(z)$ one has

$$\xi(z) = (a_j(f(z))|\varphi_j(f(z))|^2)^{1/m} |J_j(z)|^2.$$

For $0 < t < r$, $\xi(z)/\mu_t(z)$ is continuous on $\Delta_t \cup \partial \Delta_t$ and takes 0 on $\partial \Delta = t$ and at the points where $\varphi(z) = 0$.

Let z_0 be a point where $\xi(z)/\mu_t(z)$ takes its maximum. Then $\xi(z)/\mu_t(z)$ is C^∞ at z_0 and satisfies

$$[\sqrt{-1}\partial\bar{\partial} \log \xi - \sqrt{-1}\partial\bar{\partial} \log \mu_t]_{z=z_0} \le 0,$$

so that

$$(\sqrt{-1}\partial\bar{\partial} \log \xi)^n_{z=z_0} \le (\sqrt{-1}\partial\bar{\partial} \log \mu_t)^n_{z=z_0},$$

because $\sqrt{-1}\partial\bar{\partial}\log\xi = (\sqrt{-1}/m)\partial\bar{\partial}\log a_j(f(z)) = \frac{1}{m}f^*\omega$ is positive definite at z_0 (cf. (1.6)).

Rewriting the above inequality one has

$$\frac{1}{m^n}(f^*\omega^n)_{z=z^0} \leq (n+1)^n(\omega_t^n)_{z=z_0},$$

which means in view of (2.6) that

$$(a_j(f(z_0))|\varphi_j(f(z_0))|^2)^{\frac{1}{m}}|J_j(z_0)|^2 \leq M^n(n+1)^n\kappa\mu_t(z_0),$$

or $\xi(z_0) \leq m^n(n+1)^n\kappa\mu_t(z_0)$. Hence

$$\frac{\xi(z)}{\mu_t(z)} \leq \frac{\xi(z_0)}{\mu_t(z_0)} \leq m^n(n+1)^n\kappa$$

holds for any $z \in \Delta_t$.

By letting $t \to r$ we obtain

$$\xi(z) \leq m^n(n+1)^n\kappa\mu_r(z)$$

so that

$$f^*v \leq m^n(n+1)^n\kappa v_r.$$

Corollary 1. *Let W be as above and let $f : \Delta_r \to W$ be a holomorphic map satisfying $J_j(0) \neq 0$. Then*

$$r^{2n} \leq \frac{m^n(n+1)\,n\kappa}{(a_j(f(0))|\varphi_j(0))|^2)^{\frac{1}{m}}}|J_j(0)|^2.$$

Proof. Since $\xi(0) \leq m^n(n+1)^n\kappa\mu_r(0)$ holds by the proof of the theorem, the desired inequality can be seen from $\mu_r(0) = \frac{1}{r^{2n}}$.

Corollary 2. *For any holomorphic map $f : \mathbb{C}^n \to W$, the Jacobian of f is everywhere zero.*

3 Hyperbolic Measure

Let W be a complex manifold of dimension n. A subset $X \subset W$ is said to be measureable if an open covering $W = \bigcup_\lambda U_\lambda, U_\lambda \subset \mathbb{C}^n$ exists such that $X \cap U_\lambda$ are measurable sets in the sense of Lebesgue.

Definition. (Kobayashi [12]). *Given a measurable set $X \subset W$, let $\mathcal{S}$ be at most a countably many collection of pairs (f_j, Ξ_j) $(j = 1, 2, 3, \ldots)$ such that f_j are holomorphic maps from Δ_1 to W and $\Xi_j \subset \Delta_1$ are measurable sets satisfying $\bigcup_j f_j(\Xi_j) \supset X$. Then the hyperbolic measure $\eta_W(X)$ of X is defined by*

$$\eta_W(X) = \inf_{\mathcal{S}} \sum_j v_1[\Xi_j],$$

where $v_1[\Xi_j] = \int_{\Xi_j} v_1$.

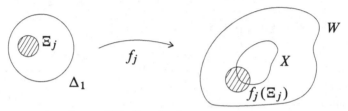

Basic properties of η_W will be listed below.

Proposition. (3.1) *η_W is countably additive, i.e. for any countably many mutually disjoint measurable set X_λ, $\lambda = 1, 2, \ldots$,*

$$\eta_W\left(\bigcup_{\lambda=1}^{\infty} X_\lambda\right) = \sum_{\lambda=1}^{\infty} \eta_W(X_\lambda)$$

holds true.

Proof. The equality is a direct consequence of the definition of η_W and the well-known countable additivity of the Lebesgue measure.

Proposition. (3.2) *Let W and V be complex manifolds of dimension n and let $f : W \to V$ be a holomorphic map. Then, for any measurable set $X \subset W$,*

$$\eta_W(X) \geq \eta_V(f(X))$$

holds.

Proof. Let $f_i : \Delta_1 \to W$ be a holomorphic map and $\Xi_j \subset \Delta_1$ be measurable such that $\bigcup_j f_j(\Xi_j) \supset X$. Then the composite $f \circ f_j : \Delta_1 \to V$ satisfies $\bigcup_j f \circ f_j(x_i_j) \supset f(X)$, so that $\bigcup_j f \circ f_j(\Xi_j) \supset f(X)$ and

$$\sum_j v_1[\Xi_j] \geq \eta_V(f(X)),$$

as a result. Hence, by taking the lower limit of the left hand side, we obtain the desired inequality.

Remark. This proposition implies that η_W depends only on the complex structure of W. Namely, for any biholomorphic map $f : W \to V$, one has $\eta_W(X) = \eta_V(f(X))$.

Proposition. (3.3) $\eta_{\Delta_r}(X) = v_r[X]$.

Proof. Since the biholomorphic map from Δ_1 to Δ_r given by $z = (z_1, \ldots, z_n) \to rz = (rz_1, \ldots, rz_n)$ induces $v = 1$ from v_r, it suffices to prove the assertion for $r = 1$. By definition

$$\eta_{\Delta_1}(X) = \inf \sum_j v'_1[\varXi_j].$$

Here the lower limit is taken with respect to holomorphic maps $f_j : \Delta_1 \to \Delta_1$ and measurable set $\varXi_j \subset \Delta_1$ such that $\bigcup_j f_j(\varXi_j) \supset X$. By Theorem (1.4) one has $f_j^* v_1 \leq v_1$. Hence

$$\sum_j v_1[\varXi_j] = \sum_j \int_{\varXi_j} v_1 \geq \sum_j \int_{\varXi_j} f_j^*(v_1) \geq \sum_j \int_{f_j(\varXi_j)} v_1$$

$$\geq \int_{\bigcup f_j(\varXi_j)} v_1 \geq \int_X v_1 = v_1[X].$$

By taking the lower limit we obtain

$$\eta_{\Delta_1}(X) \geq v_1[X].$$

Conversely, letting $f = $ identity and $\varXi = X$ one has $\eta_{\Delta_1}(X) = v_1[X]$. Therefore

$$\eta_{\Delta_1}(X) = v_1[X].$$

Proposition. (3.4) *Let $\tilde{W}$ be the universal covering space of W and let $\varpi : \tilde{W} \to W$ be the covering map. Then, for any measurable subset $X \subset \tilde{W}$ such that $\varpi : X \to \varpi(X)$ is one to one, one has $\eta_{\tilde{W}}(X) = \eta_W(\varpi(X))$.*

The proof is easy.

Proposition. (3.5) *Let W be an algebraic manifold of general type and let v be the volume form on W defined by (2.5). Then, for any $X \subset W$*

$$\eta_W(X) \geq \frac{1}{\kappa_1} \int_X v$$

holds. Here $\kappa_1 = (n+1)^n m^n \kappa$ (cf. Sect. 2).

Proof. By Theorem (2.7), $f^* v \leq \kappa_1 v_1$ holds for any holomorphic map $f : \Delta_1 \to W$. Let $f_j : \Delta_1 \to W$ be holomorphic and let $\Delta_j \subset \Delta_1$ be measurable such that $\bigcup_j f_j(\varXi_j) \supset X$. Then

$$\eta_W(X) = \inf_S \sum_j v_1[\varXi_j] = \inf_S \sum_j \int_{\varXi_j} v_1$$

$$\geq \frac{1}{\kappa_1} \inf_S \sum_j \int_{\varXi_j} f^* v \geq \frac{1}{\kappa_1} \inf_S \sum_j \int_{f(\varXi_j)} v \geq \frac{1}{\kappa_1} \int_X v.$$

Definition. (**Kobayashi** [12]). *A complex manifold W is said to be **measure hyperbolic** if $\eta_W(U) > 0$ holds for any nonempty open subset $Y \subset W$.*

For any measure hyperbolic manifold W, the following are true by the above Propositions (2.1), ..., (2.5).

(i) $\varDelta_r$ is measure hyperbolic since $\eta_{\varDelta_r}(U) = v_r[U] > 0$.

(ii) For any nonempty open set $U \subset W$ and a measurable set $X \subset W$, $\eta_W(X) \leq \eta_U(X)$.

(iii) If W is measure hyperbolic, every open subset of W is measure hyperbolic.

(iv) Any bounded domain $B \subset \mathbb{C}^n$ is measure hyperbolic.

(v) If the universal covering space $\tilde{W}$ of W is measure hyperbolic, W is also measure hyperbolic.

(vi) Algebraic manifolds of general type are measure hyperbolic.

(vii) $\eta_{\mathbb{C}^n} = 0$.

Proof. Let $X \subset \mathbb{C}^n$ be a measureable set. Let $\varXi = \varDelta_\epsilon$ and let $f(z) = rz$. Then $f(\varXi) \supset X$ for sufficiently large r, since X is bounded. Hence $\eta_{\mathbb{C}^n}(X) \leq v_1(\varXi)$. The desired equality is obtained by letting $\epsilon \to 0$.

(viii) $W = \bigcup_{\lambda=1}^{\infty} U_\lambda, \quad \eta_{U_\lambda} = 0 \ (\lambda = 1, 2, \ldots) \Rightarrow \eta_W = 0$.

Proof. $\eta_W(X) \leq \sum_\lambda \eta_W(X \cap U_\lambda) \leq \sum_\lambda \eta_{U_\lambda}(X \cap U_\lambda) = 0$.

In particular one has $\eta_{\mathbb{P}^n} = 0$.

Problem.
Find all measure hyperbolic compact complex manifolds.
  (i) When $n = 1$, i.e. when W is a compact Riemann surface,

genus of W	universal covering	η_W
0	$\mathbb{P}_1$	0
1	$\mathbb{C}$	0
≥ 2	$\varDelta_1 (\cong \mathbb{D})$	0

In particular, if the genus of W is ≥ 2, $\eta_W(W) = \eta_{\varDelta_1}(\overline{W}) = (g-1)\pi$. Here $\overline{W}$ denotes the fundamental domain of W in $\varDelta_1$.

(ii) When $n = 2$, i.e. when W is a compact surface (= a compact complex manifold of dimension 2), let $Q : \tilde{W}_p \to W$ be the quadratic transformation at $p \in W$.

Then $Q_p^{-1}(p) = C \cong \mathbb{P}_1$ and the self-intersection number $I(C, C)$ is equal to -1.

Definition. *An exceptional curve of the 1st kind is a curve* $C \subset W$ *such that* $C \cong \mathbb{P}_1$ *and* $I(C, C) = -1$.

Theorem. *If* $C \subset W$ *is an exceptional curve of the first kind, then* C *can be contracted to a point. Namely, there exist a (nonsingular) surface* V *and* $p \in V$ *such that* $W = Q_p(V)$.

Definition. W *is said to be minimal if* W *does not contain any exceptional curve of the first kind.*

Theorem. (3.6) *For any compact complex surface* W, *there exists a minimal surface* W_0 *such that* $W = Q_{p_1} Q_{p_2} \cdots Q_{p_n}(W_0)$. W_0 *is called a minimal model of* W.

As for the minimal surfaces, one has the following, according to the classification theory of Kodaira [14].

classification	b_1	surfaces	η_W
I_0	even	$\mathbb{P}_2$, ruled surfaces	0
II_0	even	K3-surfaces	?
III_0	even	tori	0
IV_0	even	elliptic surfaces	0
V_0	even	algebraic surfaces of general type	measure hyperbolic
VI_0	odd	elliptic surfaces	0
VII_0	odd	?	?

If W is an elliptic surface, there exists a compact Riemann surface Γ and a surjective holomorphic map $\Phi : W \to \Gamma$, by definition, such that $\Phi^{-1}(w)$ are elliptic curves for $w \neq a_\lambda (\lambda = 1, \ldots, q)$. $\eta_W = 0$ follows from this assertion.

By Theorem (3.6), any compact surface can be obtained by a succession of quadratic transformations.

Therefore we obtain:

classification	η_W	classification	η_W
I	0	IV	0
II	?	V	measure hyperbolic
III	?	VI	0
		VII	?

(3.7) *Problem 1.* What is the relation between $\eta_{Q_p(W)}$ and η_W?

(3.8) *Problem 2.* Given an analytic family $\{W_t; |t| < 1 \mid W_0 = W\}$ of W, is $\eta_{W_t}(W_t)$ a continuous function of t?

Example 1. Let W be a $K3$ surface. Then there exists an analytic family $\{W_t; |t| < 1 \; t \in \mathbb{C}^{20}\}$, $W_0 = W$, such that W_t are elliptic surfaces for $0 < |t| < 1$. For such a family $\eta_{W_t} = 0$ holds. In general nothing is known.

Example 2. If W is a compact Riemann surface of genus ≥ 2, $\eta_{W_t}(W_t) = (g-1)\pi$.

Let $\mathbb{D}$ be the polydisc $\{x \in \mathbb{C}^n \mid |z_\alpha| < 1\}$. Recall that the volume form $\tilde{v}_1$ was defined by

$$\tilde{v}_1 = \prod_{\alpha=1}^n \frac{1}{(1 - |z_\alpha|^2)^2} \, dV(z) \quad \text{(cf. §6 Example 1).}$$

Proposition. *For any measurable set $X \subset \mathbb{D}$,*

$$\eta_{\mathbb{D}}(X) = \tilde{v}_1[X].$$

Proof. $\eta_{\mathbb{D}}$ and $\tilde{v}_1$ are invariant under biholomorphic maps. Hence $\eta_{\mathbb{D}}(X) = c\tilde{v}_1[X]$ for some constant c. For the natural embedding $\iota : \Delta_1 \to \mathbb{D}$, $\eta_{\mathbb{D}}(\iota(X)) \leq \eta_{\Delta_1}(X))$ holds for any $X \subset \Delta_1$ by Proposition (3.2).

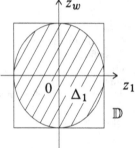

Namely, $c\tilde{v}_1[X] \leq v_1[X]$, whence $c\tilde{v}_1 \leq v_1$ follows. Since $\tilde{v}_1(0) = v_1(0)$, c is at most 1. We shall show that $f^*\tilde{v}_1 \leq v_1$ holds for any holomorphic map $f : \Delta_1 \to \mathbb{D}$, $f(z) = (f_1(z), \ldots, f_n(z)) \in \mathbb{D})$. We may assume that $f(0) = 0$, so that it suffices to prove that the modulus of the Jacobian

$$|J_f(0)| = \left| \det \left(\frac{\partial f_\alpha}{\partial z_\beta} \right)_{z=0} \right|$$

does not exceed 1.

Let

$$f_\alpha(z) = c_{\alpha 1} z_1 + \cdots + c_{\alpha n} z_n + \cdots$$

be the Taylor series.

By assumption,

$$|f_\alpha(t z_1, \ldots, t z_n)| < 1)$$

if $|z| = 1$ and $|t| < 1$. By the Schwarz Lemma (3.1),

$$\left|\frac{df_\alpha}{dt}(tz_1, \ldots, tz_n)_{t=0}\right| < 1.$$

Hence $|c_{\alpha a}z_1 + \cdots + c_{\alpha n}z_n| \leq 1$ if $|z| \leq 1$, so that one has $|c_{\alpha 1}|^2 + \cdots + |c_{\alpha n}|^2 \leq 1$.
Therefore

$$|J_f(0)| = \begin{Vmatrix} c_{11} & \ldots & c_{1n} \\ . & . & . \\ c_{n1} & \ldots & c_{nn} \end{Vmatrix} \leq 1.$$

Given $X \subset \mathbb{D}$, let $\mathcal{S}$ be the set of $\{f_j, \varXi_j; j = 1, 2, \ldots\}$ such that $f_j : \varDelta_1 \to \mathbb{D}$ are holomorphic maps and $\varXi_j$ are subsets of $\varDelta_1$ such that $X \subset \bigcup_j f_j(\varXi_j)$.
Since

$$\sum_j \int_{Xi_j} f^* \tilde{v}_1 \geq \sum_j \int_{f_j(\varXi_j)} \tilde{v}_1 \geq \int_X \tilde{v}_1 = \tilde{v}_1[X],$$

by taking the lower limit one has

$$\eta_{\mathbb{D}}(X) = \inf_{\mathcal{S}} \sum_j \int_{\varXi_j} \tilde{v}_1 \geq \tilde{v}_1[X].$$

Therefore $\eta_{\mathbb{D}}(X) = c\tilde{v}_1[X]$ with $c \geq 1$, so that $c = 1$ since one has already $c \leq 1$.

Definition.

$$\eta_W^{\mathbb{D}}(X) = \inf_{\mathcal{S}} \sum_j \tilde{v}_1[\varXi_j].$$

Here the lower limit is taken over the set $\mathcal{S}$ of those $(f_j, \varXi_j), (j = 1, 2, \ldots)$ satisfying:
(i) $\varXi_j \subset \mathbb{D}$ is measurable.
(ii) $f_j : \mathbb{D} \to W$ is a holomorphic map.
(iii) $\bigcup_j f_j(\varXi_j) \supset X$.

Let us denote the original η_W by η_W^δ. Then, for any $X \subset W$,

$$1 \leq \frac{\eta_W^\delta(X)}{\eta_W^{\mathbb{D}}(X)} \leq n^n.$$

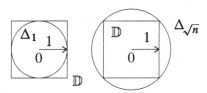

Consequently, being measure hyperbolic with respect to Δ and $\mathbb{D}$ are equivalent to each other.

$$\begin{cases} \eta_\Delta^\Delta(X) = v_1[X] \\ \eta_\mathbb{D}^\mathbb{D}(X) = \tilde{v}_1[X] \\ \eta_\mathbb{D}^\Delta(X) = \tilde{v}_1[X]. \end{cases}$$

$$\begin{cases} \dfrac{\eta_\mathbb{D}^\mathbb{D}(X)}{\eta_\mathbb{D}^\Delta(X)} = 1 \\ \dfrac{\eta_\Delta^\mathbb{D}(X)}{\eta_\Delta^\Delta(X)} = ? \end{cases}$$

Problem. *Does* $\dfrac{\eta_W^\mathbb{D}(X)}{\eta_W^\Delta(X)}$ *depend on X and W?*

Chapter 3
Nevanlinna Theory of One Variable (2)

Abstract The Nevanlinna theory as stated in Chap. 1 is reformulated in such a way that it is extendable to several variables.

Keywords The second main theorem · Defect relation

The Nevanlinna theory of one variable, as was stated in Chap. 1, will be reformulated below so that it can be extendable to the case of several variables (cf. [4]).

1 Holomorphic Maps to the Riemann Sphere

As in Sect. 2, we shall use the following notation:

Δ_r : the disc of radius $r = \{z \in \mathbb{C} \mid |z| < r\}$,
$\partial \Delta_r$: the boundary of $\Delta_r = \{z \in \mathbb{C} \mid |z| = r\}$,
$\mathbb{P}_1$: the Riemann sphere $= \mathbb{C} \cup \{\infty\}$.

For any meromorphic function f on $\mathbb{C}$, the distribution of the roots of $f(z) - a = 0$ $(a \in \mathbb{P}_1)$ will be described below.

Definition (cf. Sect. 4).

$$n(r, a) = \{\textit{the number of roots of } f(z) - a = 0 \textit{ in } \Delta_r,$$
$$\textit{counted with multiplicity}\}.$$

Problem. What is the growth rate of $n(r, a)$ for $r \to \infty$?

© The Author(s) 2017
K. Kodaira, *Nevanlinna Theory*, SpringerBriefs in Mathematics,
https://doi.org/10.1007/978-981-10-6787-7_3

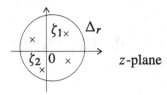

Definition.

$$N(r, a) = \int_0^r n(t, a)\frac{dt}{2\pi t}.$$

$N(r, a) < \infty$ if $f(0) \neq a$.

Recall that the canonical Kähler metric is

$$\omega = \frac{\sqrt{-1}}{2\pi}\frac{dw \wedge d\overline{w}}{(1 + |w|^2)^2} \quad (w \in \mathbb{C}, \quad \mathbb{P}_1 = \mathbb{C} \cup \{\infty\}) \quad \text{(cf. Sect. 1).}$$

Remark. $\int_{\mathbb{P}_1} \omega = 1.$

Identifying a meromorphic function f as a holomorphic map $f : \mathbb{C} \to \mathbb{P}_1$, the pull-back $f^*\omega$ of ω by f is expressed as

$$f^*\omega = \frac{\sqrt{-1}}{2\pi}\frac{|f'(z)|^2}{(1 + |f(z)^2)^2} dz \wedge d\overline{z},$$

which we shall write as $\xi(z)\frac{\sqrt{-1}}{2} dz \wedge d\overline{z}$, where

$$\xi(z) = \frac{|f'(z)|^2}{\pi(1 + |f(z)|^2)^2}.$$

Definition.

$$\begin{cases} A(t) = \displaystyle\int_{\Delta_t} f^*\omega \\[2mm] T(r) = \displaystyle\int_0^r A(t)\frac{dt}{2\pi t}. \end{cases}$$

Definition. *The average of a function g over $\partial\Delta_r$ is*

$$\mathcal{M}_r(g) = \frac{1}{2\pi}\int_0^{2\pi} g(re^{i\theta}) d\theta.$$

Definition.

$$
\begin{cases}
u_a(z) = \dfrac{1}{4\pi} \log \dfrac{(1 + |f(z)|^2)(1 + |a|^2)}{|f(z) - a|^2} \\
m(r, a) = \mathcal{M}_r(u_a).
\end{cases}
$$

2 The First Main Theorem

Theorem. (2.1) **(The first main theorem; Theorem (4.3))**

$$
T(r) = N(r, a) + m(r, a) - m(0, a).
$$

Proof. We put $d^\perp = \sqrt{-1}(\bar{\partial} - \partial)$ so that $dd^\perp = 2\sqrt{-1}\partial\bar{\partial}$. By the equality (4.1), $dd^\perp u_a = f^*\omega$. Hence

$$
A(t) = \int_{\Delta_t} dd^\perp u_a(z).
$$

By setting $\tau(t) = \frac{1}{2\pi} \log t$, or $d\tau = \frac{dt}{2\pi t}$, we have

$$
T(r) = \iint_{|z|<t<r} d\tau \wedge dd^\perp u_a(z) = \iint_{\mathcal{C}} d(\tau \wedge dd^\perp u_a(z)),
$$

where

$$
\mathcal{C} = \{(t, z) \in \mathbb{R} \times \mathbb{C} \mid |z| < t < r\}.
$$

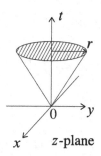

Since

$$
\partial\mathcal{C} = \Delta_r \cup \{|z| = t < t\},
$$

applying Stokes's theorem one has

$$T(r) = \int_{\Delta_r} \tau(r) dd^\perp u_a - \int_{\Delta_r} \tau(|z|) dd^\perp u_a = \int_{\Delta_r} [\tau(r) - \tau(|z|)] dd^\perp u_a(z)$$

$$(2.2) \qquad = \int_{\Delta_r} d([\tau(r) - \tau(|z|)] \wedge d^\perp u_a) + \int_{\Delta_r} d\tau(|z|) \wedge d^\perp u_a.$$

Since $d\tau(|z|) \wedge d^\perp u_a = du_a \wedge d^\perp \tau(|z|)$, the second term in (2.2) becomes

$$\int_{\Delta_r} du_a \wedge d^\perp \tau(|z|) = \int_{\Delta_r} d(u_a d^\perp \tau(|z|)) \qquad (\text{since } dd^\perp \tau(|z|) = 0)$$

$$= \lim_{\varepsilon \to 0} \int_{\Delta \backslash \Delta_\varepsilon} d(u_a d^\perp \tau) = \lim_{\varepsilon \to 0} \int_{\partial \Delta_r \backslash \partial \Delta_\varepsilon} u_a d^\perp \tau$$

$$= m(r, a) - m(0, a) \quad \left(\text{since } d^\perp \tau = \frac{d\theta}{2\pi} \right).$$

Let ζ_h, $h = 1, 2, \ldots$ be the roots of $f(z) - a = 0$ and choose $R > r$ so that

$$f(z) - a = \prod_{|\zeta_h| < R} (z - \zeta_h) g(z),$$

where $g(z)$ is a meromorphic function satisfying $g(z) \neq 0$ on Δ_R (cf. the proof of Lemma (4.2)).

Hence, for any $z \in \Delta_r$,

$$u_a(z) = \frac{1}{4\pi} \sum_{|\zeta_h| < R} \log \frac{1}{|z - \zeta_h|^2} + \{C^\infty \text{ function}\}.$$

Therefore, the first term of (2.2) is

$$= \int_{\partial \Delta_r} [\tau(r) - \tau(|z|)] d^\perp u_a - \sum_{\zeta_h \in \Delta_r} \oint_{\zeta_h} [\tau(r) - \tau(|z|)] d^\perp u_a$$

$$= - \sum_{\zeta_h \in \Delta_r} \oint_{\zeta_h} [\tau(r) - \tau(|z|)] d^\perp u_a$$

$$= - \sum_{\zeta_h \in \Delta_r} \lim_{\varepsilon \to 0} \int_{|z - \zeta_h| = \varepsilon} [\tau(r) - \tau(|z|)] d^\perp \log |z - \zeta_h|$$

$$= \sum_{\zeta_h \in \Delta_r} \frac{1}{2\pi} \lim_{\varepsilon \to 0} \int_0^{2\pi} [\tau(r) - \tau(|\zeta_h + \varepsilon e^{i\theta}|)] d\theta = \sum_{\zeta_h \in \Delta_r} [\tau(r) - \tau(|\zeta_h|)]$$

$$= \int_0^r n(t, a) d\tau(t) = N(r, a).$$

As a result

$$T(r) = N(r, a) + m(r, a) - m(0, a).$$

Definition. (cf. Sect. 2)

$$n_1(t) = \{the\ number\ of\ roots\ of\ f'(z) = 0\ in\ \Delta_t$$
$$counted\ with\ multiplicity\}.$$

$$N_1(r) = \int_0^r n_1(t) \frac{dt}{2\pi t}.$$

($N_1(r) < \infty$ if $f'(0) \neq 0$).

Remark. When $f(z) = \infty$, i.e. when z is a pole of f, letting $\frac{1}{f(z)} = g(z)$ we count the roots of $g'(z) = 0$. Note that

$$\xi(z) = \frac{|f'|^2}{\pi(1 + |f|^2)^2} = \frac{|g'/g^2|^2}{\pi(1 + \frac{1}{|g|^2})^2} = \frac{|g'|^2}{\pi(1 + |g|^2)^2}.$$

Definition. $M(r) = \mathcal{M}_r(\frac{1}{4\pi} \log \xi)$.

Similarly to the proof of the first main theorem (2.1),

(2.4) $$- 2T(r) = -N_1(r) + M(r) - M(0).$$

Remark. Let $B(t)$ be as in Sect. 2. Then

$$B(t) = -\frac{1}{2} \int_{\Delta_t} Rf^*\omega = -2A(t) \quad (since\ R = 4).$$

Hence (2.4) is equivalent to Theorem (2.4).

3 The Second Main Theorem

In this section we shall prove the following.

Theorem. (3.1) (**The second main theorem**) *Let $f(z)$ be a nonconstant meromorphic function on $\mathbb{C}$. Then, for any $a_1, \ldots, a_q \in \mathbb{P}_1$, the following inequality holds for $r \to +\infty, r \neq E$:*

$$\sum_{\lambda=1}^{q} m(r, a_\lambda) + N_1(r) \leq 2T(r) + O(\log T(r)).$$

Here E is a measurable subset of $\{r \geq 0\}$ such that $\int_E d(r^\beta) < +\infty$ is satisfied for some $0 < \beta < 1$.

For the proof of the theorem, we introduce the quantities δ and ρ_a. ρ_a will play a particularly important role.

Definition.

$$\delta(w, a)^2 = \frac{|w - a|^2}{e^4(1 + |w|^2)(1 + |a|^2)}$$

$$\rho_a(w) = \frac{\kappa}{[\log \delta^2]^2 \delta^2},$$

where $\kappa > 0$ is a constant.

Remark. (3.2) $e^4 = 54.598 \cdots$. Since δ is $\frac{1}{e^2}$ times the distance between $\hat{w}$ and $\hat{a}$, $\delta^2 \leq \frac{1}{e^4}$, so that

$$\log \frac{1}{\delta^2} \geq 4$$

and

$$[\log \delta^2]^2 = \left[\log \frac{1}{\delta^2}\right]^2 \geq 16.$$

$\rho_a(w)$ is a C^∞ function for $w \neq a$ and satisfies

$$\rho_a \leq O\left(\frac{1}{(\log |w - a|^2)^2 |w - a|^2}\right)$$

on a neighborhood of a.

Lemma. (3.3) *If* $0 < \kappa \leq \frac{2}{e^4}$,

$$\frac{1}{4\pi} dd^\perp \log \rho_a(w) \geq \rho_a(w)\omega.$$

Proof. By the definition of ρ_a,

$$\partial \bar{\partial} \log \rho_a$$

$$= \left(1 - \frac{2}{\log \frac{1}{\delta^2}}\right) \partial \bar{\partial} \log \frac{1}{\delta^2} + \frac{2}{(\log \frac{1}{\delta^2})^2} \partial \log \frac{1}{\delta^2} \wedge \bar{\partial} \log \frac{1}{\delta^2}.$$

By the definition of δ,

$$\partial \bar{\partial} \log \frac{1}{\delta^2} = \frac{dw \wedge d\bar{w}}{(1 + |w|^2)^2}$$

and

$$\partial \log \frac{1}{\delta^2} = -\frac{1 + a\bar{w}}{(1 + |w|^2)(w - a)} dw.$$

Hence

$$\partial\bar{\partial}\log\rho_a = \left(1 - \frac{2}{\log\frac{1}{\delta^2}} + \frac{2|1+a\overline{w}|^2}{[\log\frac{1}{\delta^2}]^2|w-a|^2}\right)\frac{dw\wedge d\overline{w}}{(1+|w|^2)^2}$$

$$\geq \left(\frac{1}{2} + \frac{2|1+a\overline{w}|^2}{[\log\frac{1}{\delta^2}]^2|w-a|^2}\right)\frac{dw\wedge d\overline{w}}{(1+|w|^2)^2}$$

$$\geq \frac{4|w-a|^2 + 4|1+a\overline{w}|^2}{2[\log\delta^2]^2|w-a|^2}\frac{dw\wedge d\overline{w}}{(1+|w|^2)^2}$$

$$= \frac{2(1+|w|^2)(1+|a|^2)}{[\log\delta^2]^2|w-a|^2}\frac{dw\wedge d\overline{w}}{(1+|w|^2)^2} = \frac{2}{\kappa e^4}\rho_a(w)\frac{dw\wedge d\overline{w}}{(1+|w|^2)^2}.$$

Therefore, for $0 < \kappa \leq \frac{2}{e^4}$,

$$\frac{1}{4\pi}dd^\perp\log\rho_a = \frac{1}{2\pi}\partial\bar{\partial}\log\rho_a \geq \rho_a(w)\omega.$$

Lemma. (3.4) $\int_{\mathbb{P}_1} dd^\perp\log\rho_a < +\infty$.

Proof. $dd^\perp\log\rho_a$ is a C^∞ form for $w \neq a$ and

$$dd^\perp\log\rho_a = O\left(\frac{1}{[\log|w-a|^2]^2|w-a|^2}\omega\right)$$

on a neighborhood of a. Hence, it suffices to show that the integral

$$\int_{|w|<\varepsilon}\frac{\frac{\sqrt{-1}}{2}dw\wedge d\overline{w}}{[\log|w|^2]^2|w|^2}$$

is finite for sufficiently small ε. But this integral is

$$= \pi\int_0^\varepsilon\frac{dr^2}{[\log r^2]^2 r^2} = \frac{\pi}{\log\frac{1}{\varepsilon^2}} < +\infty.$$

Definition.

$$\begin{cases} \tilde{u}_a(z) = \dfrac{1}{4\pi}\log\rho_a(f(z)) \\ \mathcal{L}_a(a) = \dfrac{1}{2\pi}\log\log\dfrac{1}{\delta^2}. \end{cases}$$

Since $u_a(z) = \frac{1}{4\pi}\log\frac{1}{\delta^2} - \frac{1}{\pi}$,

$$\tilde{u}_a = \frac{1}{4\pi}\left\{\log\frac{1}{\delta^2} - 2\log\log\frac{1}{\delta^2} + \log\kappa\right\} = u_a - \mathcal{L}_a + +\frac{1}{4\pi}(4+\log\kappa).$$

$$(3.5) \quad = u_a - \mathcal{L}_a + \kappa_1 \quad \left(\kappa_1 = \frac{1}{4\pi}(4+\log\kappa)\right).$$

We shall use the following notation.
(3.6) Notation

$$\begin{cases} \tilde{A}_a(t) = \displaystyle\int_{\Delta_t} dd^{\perp}\tilde{u}_a(z) \\ \tilde{T}_a(r) = \displaystyle\int_0^r \tilde{A}_a(t)\frac{dt}{2\pi t} \\ \tilde{m}(r, a) = \mathcal{M}_r(\tilde{u}_a) \\ \mu_a(r) = \mathcal{M}_r(\mathcal{L}_a). \end{cases}$$

Averaging both sides of (3.5) over $\partial\Delta_r$, one has

(3.7) $\tilde{m}(r, a) = m(r, a) - \mu_a(r) + \kappa_1.$

One has also a relation

(3.8) $\dfrac{1}{2\pi}\log 4 \le \mu_a(r) \le \dfrac{1}{2\pi}\log\left(m(r, a) + \dfrac{1}{\pi}\right) + \dfrac{1}{2\pi}\log 4\pi.$

In fact,

$$\mu_a(r) = \mathcal{M}_r\left(\frac{1}{2\pi}\log\log\frac{1}{\delta^2}\right)$$

$$\le \frac{1}{2\pi}\log\mathcal{M}_r\left(\log\frac{1}{\delta^2}\right) \qquad \text{(since log is concave)}$$

$$= \frac{1}{2\pi}\log\{4\pi m(r, a) + 4\} = \frac{1}{2\pi}\log\left(m(r, a) + \frac{1}{\pi}\right) + \frac{1}{2\pi}\log 4\pi$$

and, since $\log\frac{1}{\delta^2} \ge 4$,

$$\mu_a(r) \ge \mathcal{M}_r\left(\frac{1}{2\pi}\log 4\right) = \frac{1}{2\pi}\log 4.$$

Lemma. (3.9) $\tilde{T}_a(r) = T(r) - \mu_a(r) + \mu_a(0).$

Proof. By the equality (3.5),

$$dd^{\perp}\tilde{u}_a = dd^{\perp}u_a - dd^{\perp}\mathcal{L}_a.$$

Integration on Δ_t gives

$$\int_{\Delta_t} dd^{\perp}\tilde{u}_a(z) = \int_{\Delta_t} dd^{\perp}u_a(z) - \int_{\Delta_t} dd^{\perp}\mathcal{L}_a(z).$$

Hence

$$\tilde{A}_a(t) = A(t) - \int_{\Delta_t} dd^\perp \mathcal{L}_a(z)$$

and

$$\tilde{T}_a(r) = T(r) - \int_0^r \frac{dt}{2\pi t} \int_{\Delta_t} dd^\perp \mathcal{L}_a(z).$$

Putting $\tau(t) = \frac{1}{2\pi} \log t$, we express the second term, say I, of the right hand side of the above equality as

$$I = \iint_{|z|<t<r} d\tau(t) \wedge dd^\perp \mathcal{L}_a(z) = \iint_{|z|<t<r} d(\tau(t) dd^\perp \mathcal{L}_a(z)).$$

Let the roots of the equation $f(z) - a = 0$ be denoted by $\zeta_1, \cdots, \zeta_h, \cdots$. Then, $dd^\perp \mathcal{L}_a(z)$ is C^∞ for $z \neq \zeta_h$. We put

$$\Delta_{r,\varepsilon} = \Delta_r \setminus \bigcup_{h=1,2,\ldots} \{z \mid |z - \zeta_h| \leq \varepsilon\}.$$

Then, by Stokes's theorem

$$I = \lim_{\varepsilon \to 0} \int_{\Delta_{r,\varepsilon}} [\tau(r) - \tau(|z|)] dd^\perp \mathcal{L}_a.$$

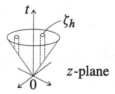

Hence

(3.10) $$I = \int_{\Delta_r} d([\tau(r) - \tau(|z|)] d^\perp \mathcal{L}_a) + \int_{\Delta_r} d\tau(|z|) \wedge d^\perp \mathcal{L}_a(z).$$

By the definition of $\mathcal{L}_a$,

$$d^\perp \mathcal{L}_a = \frac{1}{2\pi} \frac{1}{\log \frac{1}{\delta^2}} d^\perp \log \frac{1}{\delta^2} = \frac{2}{\log \frac{1}{\delta^2}} d^\perp u_a(z).$$

Letting z tend to ζ_h, one has $\delta \to 0$, so that

$$\frac{1}{\frac{1}{\log \delta^2}} \to 0.$$

Hence

$$\lim_{\varepsilon \to 0} \int_{|z-\zeta_h|=\varepsilon} \frac{1}{\frac{1}{\delta^2}} d^\perp u_a(z) = 0 \qquad \left(\text{since } \oint_{\zeta_h} d^\perp u_a = 1, \quad \text{cf. Sect. } 4\right).$$

Hence the first term of (3.10) becomes

$$= \int_{\partial \Delta_r} [\tau(r) - \tau(|z|)] d^\perp \mathcal{L}_a - \sum_{\zeta_h \in \Delta_r} \lim_{\varepsilon \to 0} \int_{|z-\zeta_h|=\varepsilon} [\tau(r) - \tau(|z|)] \frac{2}{\log \frac{1}{\delta^2}} d^\perp u_a = 0.$$

Therefore, by exploiting $d\tau \wedge d^\perp \mathcal{L}_a = d\mathcal{L}_a \wedge d^\perp \tau$, one deduces from (3.10) that

$$I = \int_{\Delta_r} d\mathcal{L}_a \wedge d^\perp \tau(|z|)$$

$$= \int_{\Delta_r} d(\mathcal{L}_a d^\perp \tau) \quad (\text{since } dd^\perp \tau = 0) = \lim_{\varepsilon \to 0} \int_{\Delta_r \setminus \Delta_\varepsilon} d(\mathcal{L}_a d^\perp \tau)$$

$$\int_{\partial \Delta_r} \mathcal{L}_a d^\perp \tau(|z|) - \lim_{\varepsilon \to 0} \mathcal{L}_a d^\perp \tau(|z|) = \mu_a(r) - \mu_a(0).$$

Proof of Theorem (3.1) Let $\rho(w) = \prod_{\lambda=1}^q \rho_{a_\lambda}(w)$. Then $\rho(w)$ is C^∞ for $w \neq a_\lambda$ and $\rho > 0$.

Lemma. (3.11) $\int_{\mathbb{P}_1} \rho(w)\omega < +\infty$.

Proof. On a neighborhood of $w = a_\lambda$,

$$\rho(w) \leq O\left(\frac{1}{[\log|w - a_\lambda|]^2 |w - a_\lambda|^2}\right).$$

so that its integral over $\mathbb{P}_1$ is finite as well as in Lemma (3.4).

Definition.

$$\begin{cases} \Psi(r) = \displaystyle\int_{\Delta_r} f^*(\rho\omega) \\ \Phi(t) = \mathcal{M}_t(\rho(f(z))\xi(z)) \\ Q(r) = \displaystyle\int_0^r \Psi(t)\frac{dt}{2\pi t}. \end{cases}$$

Let us calculate $\Psi(r)$.

$$\Psi(r) = \int_{\Delta_r} \rho(f(z))\xi(z)\frac{\sqrt{-1}}{2} dz \wedge d\bar{z}$$
$$= \int_0^r \int_0^{2\pi} \rho(f(te^{i\theta}))\xi(te^{i\theta})t\, dt\, d\theta = \int_0^r 2\pi\Phi(t)t\, dt.$$
$$\Psi(r) = 2\pi \int_0^{2\pi} \phi(t)t\, dt.$$

Basic Lemma. *Let $0 < \beta < 1$ and let $\nu = (\frac{2}{\beta}-1)^2 > 1$. Then there exists an open set $E \subset \{r \geq 0\}$ such that $\Phi(r) \leq Q(r)^\nu$ holds for any $r \notin E$.*

Proof. Let $\zeta_h(a_h)$, $h = 1, 2, \ldots$ be the roots of $f(z) - a_\lambda = 0$. Since $\rho(f(z))\xi(z)$ is C^∞ at $z \neq \zeta_h(a_h)$, $\Phi(t)$ is C^∞ for $t \neq |\zeta_h(a_h)|$.

We put $E = \{r \mid \Phi(r) > Q(r)^\nu\}$.
For simplicity we put $\hat{\Psi}(t) = \frac{1}{2\pi}\Psi(t)$, so that $Q(r) = \int_0^r \hat{\Psi}(t)\, dt$. Then, by letting $\lambda = \sqrt{\nu} = \frac{2}{\beta} - 1$, it turns out that at least one of the following is true for $r \in E$:

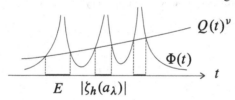

(i) $\Phi(r) > r^{\beta-2}\hat{\Psi}(r)^\lambda$,
(ii) $\hat{\Psi}(r) > r^\beta Q(r)^\lambda$.

In fact, if neither of these were true one would have

$$\Phi(r) \leq r^{\beta-2}(r^\beta Q(r)^\lambda)^\lambda = Q(r)^\lambda$$

which contradicts $r \in E$.

Hence, for each $r \in E$ one has either

(i) $dr^\beta = \beta r^{\beta-1}\, dr = \beta r^{\beta-2}\frac{d\hat{\Psi}}{\Phi} < \beta\frac{d\hat{\Psi}}{\hat{\Psi}^\lambda}$ (note that $d\hat{\Psi} = \Phi r\, dr$)
or
(ii) $dr^\beta = \beta r^\beta \frac{dQ}{\hat{\Psi}} < \beta\frac{dQ}{Q^\lambda}$ (note that $r\, dQ = \hat{\Psi}\, dr$).

Therefore

$$dr^\beta < \beta \left(\frac{d\hat{\Psi}(r)}{\hat{\Psi}(r)^\lambda} + \frac{dQ(r)}{Q(r)^\lambda} \right).$$

Hence

$$\int_{E\cap\{r\geq 1\}} dr^\beta < \beta \int_0^\infty \frac{d\hat{\Psi}(r)}{\hat{\Psi}(r)^\lambda} + \frac{dQ(r)}{Q(r)^\lambda}$$

$$\leq \frac{\beta}{\lambda-1} \left(\frac{1}{\hat{\Psi}(1)^{\lambda-1}} + \frac{1}{Q(1)^{\lambda-1}} \right) < +\infty,$$

so that $\int_E dr^\beta < +\infty$.

Remark.

$$Q(r) = \int_0^r \frac{dt}{t} \int_0^t \Phi(s)s\,ds = \int_0^r [\log r - \log s]\Phi(s)s\,ds.$$

Corollary. *Let $Q(r)$ be as above defined for a continuous function $\Phi(s)$ on $[0, \infty)$. Then*

$$\int_E d(r^\beta) < +\infty$$

if $0 < \beta < 1$ and $E = \{r \mid r \geq 0, \Phi(r) > Q^{(\frac{2}{\beta}-1)^2}\}$.

Problem. Find an alternate proof of the corollary.

For sufficiently small κ,

(3.14) $$\rho(w) \leq \sum_{\lambda=1}^q \rho_{a_\lambda}(w)$$

Combining this inequality with Lemma (3.3), one has

$$\rho(w)\omega \leq \frac{1}{4\pi} \sum_{\lambda=1}^q dd^\perp \log \rho_{a_\lambda}(w),$$

so that

$$f^*(\rho\omega) \leq \frac{1}{4\pi} \sum_{\lambda=1}^q dd^\perp \log \rho_{a_\lambda}(f(z)) = \sum_{\lambda=1}^q dd^\perp \tilde{u}_{a_\lambda}.$$

Hence, the integration on Δ_t yields

$$\Psi(t) \leq \sum_{\lambda=1}^q \tilde{A}_{a_\lambda}(t)$$

and further

(3.15)
$$Q(r) \leq \sum_{\lambda=1}^{q} \tilde{T}_{a_\lambda}(r).$$

Hence, by applying the estimate (3.8) in Lemma (3.9), we obtain:

(3.16)
$$Q(r) < qT(r) + const.$$

Therefore, by using (2.4), the equality to be proved becomes

$$\sum_{\lambda=1}^{q} m(r, a_\lambda) + M(r) = O(\log T(r)).$$

This can be modified by (3.7) as

(3.17)
$$\sum_{\lambda=1}^{q} \hat{m}(r, a_\lambda) + M(r) + \sum_{\lambda=1}^{q} \mu_{a_\lambda(r)} = O(\log T(r)).$$

On the other hand, since $N(r, a) \geq 0$,

$$T(r) \geq m(r, a_\lambda) - m(0, a_\lambda).$$

Substituting this to the estimate (3.8) we obtain

$$\mu_{a_\lambda}(r) = O(\log T(r)).$$

Hence, in view of (3.7), it suffices to prove the following:

$$\sum_{\lambda=1}^{q} \tilde{m}(r, a_\lambda) + M(r) = O(\log T(r)).$$

The left hand side of the inequality is estimated as

$$\mathcal{M}_r \left(\sum_{\lambda=1}^{q} \tilde{u}_{a_\lambda} + \frac{1}{4\pi} \log \xi \right) = \mathcal{M}_r \left(\frac{1}{4\pi} \log \prod_{\lambda=1}^{q} \rho_{a_\lambda}(f)\xi \right)$$

$$= \frac{1}{4\pi} \mathcal{M}_r (\log (\rho(f)\xi)) \leq \frac{1}{4\pi} \log \mathcal{M}_r (\rho(f)\xi)) = \frac{1}{4\pi} \log \Phi(r),$$

so that, for any $r \notin E$ the inequality (3.13) of the basic lemma implies

$$\sum_{\lambda=1}^{q} \hat{m}(r, a_\lambda) + M(r) \le \frac{\nu}{4\pi} \log Q(r) \le \frac{\nu}{4\pi} [\log (qT(r) + \text{const.})] \quad \text{(by (3.16))}.$$

Hence, for $r \to +\infty$, $r \notin E$, we obtain

$$\sum_{\lambda=1}^{q} \tilde{m}(r, a_\lambda) + M(r) = O(\log T(r)).$$

Definition (cf. Sect. 5).

$$\delta(a) = \liminf_{r \to +\infty} \frac{m(r, a)}{T(r)} \quad \text{(the defect of } f \text{ at } a).$$

Definition.

$$\delta_1 = \liminf_{r \to +\infty} \frac{N_1(r)}{T(r)}.$$

The defect relation can be obtained also from the second main theorem (3.1).

Theorem. (3.18) **(Defect relation; Theorem (5.2))**

$$\sum_{\lambda=1}^{q} \delta(a_\lambda) + \delta_1 \le 2.$$

Proof. The inequality is obtained from (3.1) by dividing it by $T(r)$ and letting $r \to +\infty$.

Example. (Nevanlinna) Let $m \in \mathbb{N}$ and

$$f(z) = \int_0^\infty e^{-\zeta^m} d\zeta.$$

Then $\delta(\infty) = 1$ because ∞ is not attained by f. Define $a_\lambda, \lambda = 1, \ldots, m$ by

$$a_\lambda = e^{\frac{2\pi i \lambda}{m}} \int_0^\infty e^{-r^m} dr.$$

Then $\delta(a_\lambda) = \frac{1}{m}$, so that

$$\sum_{\lambda=1}^{m} \delta(a_\lambda) + \delta(\infty) = 2.$$

Hence $\delta(a) = 0$ if $a \ne \infty, a_1, \ldots, a_m$ by the defect relation.

Chapter 4
Nevanlinna Theory of Several Variables

Abstract Holomorphic maps from $\mathbb{C}^n$ to compact complex manifolds are studied.

Keywords Bieberbach's example · The first main theorem · The second the-orem

Let W be a compact complex manifold of dimension n. Holomorphic maps $f :$ $\mathbb{C}^n \to W$ will be studied below (cf. Carlson–Griffiths [4]).

1 Bieberbach's Example

Bieberbach [1] constructed a biholomorphic map from $\mathbb{C}^2$ to a proper open subset of $\mathbb{C}^2$.

Let $\varphi : \mathbb{C}^2 \to \mathbb{C}^2$ be a biholomorphic map satisfying the following conditions:

(i) $\varphi(0) = 0$.
(ii) If

$$\begin{cases} \varphi_1(z) = a_{11}z_1 + a_{12}z_2 + \{higher\ order\ terms\} \\ \varphi_2(z) = a_{21}z_1 + a_{22}z_2 + \{higher\ order\ terms\} \end{cases}$$

is the power series expansion of $\varphi(z) = (\varphi_1(z), \varphi_2(z))$, the eigenvalues λ, μ of the matrix (a_{ij}) satisfy

$$\begin{cases} 0 < |\lambda| < 1 \\ 0 < |\mu| < 1 \\ \lambda \neq \mu^m, \lambda^m \neq \mu \quad \text{for } m = 1, 2, 3, \dots \end{cases}$$

Theorem. (**Lattés** [16], **Sternberg** [21]) *Let φ be a biholomorphic map from $\mathbb{C}^2$ to $\mathbb{C}^2$ as above. Then, there exist a neighborhood $U \subset \mathbb{C}^2$ of 0 and a biholomorphic map g from U to U such that*

$$g^{-1} \circ \varphi \circ g(z_1, z_2) = (\lambda z_1, \mu z_2).$$

© The Author(s) 2017
K. Kodaira, *Nevanlinna Theory*, SpringerBriefs in Mathematics,
https://doi.org/10.1007/978-981-10-6787-7_4

We put $\psi = g^{-1} \circ \varphi \circ g$. Then $g = \varphi^{-1} \circ g \circ \psi$ holds on $\psi^{-1}(U) \cap U$. Moreover, there exists a neighborhood $V_m \ni 0$ such that $g = \varphi^{-m} \circ g \circ \psi^m$ holds on V_m.

Definition. Let g_m be a biholomorphic map from $\psi^{-m}(U)$ to $\varphi^{-m}(g(U))$ defined by

$$g_m = \varphi^{-m} \circ g \circ \psi^m.$$

$g = g_m$ holds on V_m, so that g_m is an analytic continuation of g. Since $\bigcup_m \psi^{-m}(U) = \mathbb{C}^2$, g is extended to the whole space $\mathbb{C}^2$ analytically.

Definition. $f = \lim_{m \to \infty} g_m$.
  Hence we have a biholomorphic map $f : \mathbb{C}^2 \to B := f(\mathbb{C}^2)$.

Lemma. $B = \{z \in \mathbb{C}^2 \mid \lim_{m \to \infty} \varphi^m(z) = 0\}$.

Proof. Let $z \in B$. Then $f(w) = z$ for some $w \in \mathbb{C}^2$. If $w \in \psi^{-m}(U)$, then $f = \varphi^{-m} \circ f \circ \psi^m$ holds on $\psi^{-m}(U)$, for f is an analytic continuation of g. Hence $z = f(w) = \varphi^{-m} \circ f(\psi^m(w))$, so that $\lim_{m \to \infty} \varphi^m(z) = \lim_{m \to \infty} f(\psi^m(w)) = f(0) = 0$. Conversely, assume that $\varphi^m(z) \to 0$. Since B is a neighborhood of 0, $\varphi^m(z) \in B = f(\mathbb{C}^2)$ for sufficiently large m. Hence $\varphi^m(z) = f(w)$ holds for some $w \in \mathbb{C}^2$. Thus we obtain $z = \varphi^{-m}(f(w)) = f \circ \psi^{-m}(w) \in B$.
  In summary we have:

Proposition. (Bieberbach [1]) *Let φ be a biholomorphic map from $\mathbb{C}^2$ to $\mathbb{C}^2$ satisfying the above conditions (i) and (ii). Then $\mathbb{C}^2$ is biholomorphic to $B = \{z \in \mathbb{C}^2 \mid \lim_{m \to +\infty} \varphi^m(z) = 0\}$.*

Example. Let φ be given as follows:

$$\varphi : z = (z_1, z_2) \to (z_2, \lambda^2 z_1 + (\lambda^2 - 1)(\sin z_2 - z_2)) = \varphi(z).$$

Here $0 < |\lambda| < 1$. Then

$$(a_{ij}) = \begin{pmatrix} 0 & 1 \\ \lambda^2 & 0 \end{pmatrix},$$

so that its eigenvalues are $\pm \lambda$ and φ satisfies the conditions (i) and (ii).
  Hence, by the above proposition, one has a biholomorphic map $f : \mathbb{C}^2 \to B$ for $B = \{z \in \mathbb{C}^2 \mid \lim_{m \to +\infty} \varphi^m(z)\} = 0$. Let τ denote the following parallel transformation:

$$\tau : z = (z_1, z_2) \to (z_1 + 2\pi, z_2 + 2\pi) = \tau(z).$$

Then φ commutes with τ. i.e. $\varphi \circ \tau = \tau \circ \varphi$.
  In fact,

$$\tau \circ \varphi(z_1, z_2) = (z_2 + 2\pi, \lambda^2 z_1 + (\lambda^2 - 1)(\sin z_2 - z_2) + 2\pi) = \varphi \circ \tau(z_1, z_2).$$

Let us consider the set $B_k = \tau^k(B)$. Then

$$B_k = \{z \in \mathbb{C}^2 \mid \lim_{m \to +\infty} \varphi^m(\tau^{-k}z) = 0\} = \{z \in \mathbb{C}^2 \mid \lim_{m \to +\infty} \varphi^m(z) = \tau^k(0)\}.$$

Hence $B_k \cap B_j = \varnothing$ $(j \neq k)$. We put $f_k = \tau^k \circ f$. Then $f_k : \mathbb{C}^2 \to B_k$ is a biholomorphic map.

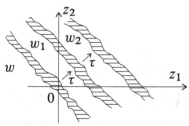

Problem. Let $\mathbb{C}^2$ be embedded as a proper subset B of $\mathbb{C}^2$ as above. Let

$$\mathbb{C}^2 = B \subset \mathbb{C}^2 = B_{(1)} \subset \mathbb{C}_1^2 = B_{(2)} \subset \mathbb{C}_2^2 = B_{(3)} \subset \cdots$$

be the repetition of this procedure and put $M = \bigcup_{\nu=1}^{+\infty} \mathbb{C}_\nu^2$.

(1) $M = \mathbb{C}^2$?   (2) Is M Stein?   (3) $H^1(M, \mathcal{O}) = ?$

2 The First Main Theorem

Let W be a compact complex manifold of dimension n and let $f : \mathbb{C}^n \to W$ be a holomorphic map. In what follows we shall assume the following.

Assumption. The Jacobian of f is not identically zero.

   Let F be a line bundle over W and let $\{f_{jk}\}$ be a 1-cocyle defining F with respect to a covering $W = \bigcup_j U_j$ (cf. Sect. 7). Let $\psi = \{\psi_j\} \in H^0(W, (F))$ be a holomorphic section of F. The divisor associated to ψ is denoted by (ψ). Let $a = \{a_j(w)\}$ be a fiber metric of F. Namely, $a_j(w)$ are positive C^∞ functions on U_j such that

$$a_j(w) = |f_{jk}(w)|^2 a_k(w)$$

is satisfied on $U_j \cap U_k$. (a is a C^∞ section of $F \otimes \overline{F}$.) Note that

$$\frac{|\psi_j(w)|^2}{a_j(w)} = \frac{|\psi_k(w)|^2}{a_k(w)}$$

holds on $U_j \cap U_k$.

Definition. $|\psi|^2(w) = \frac{\psi_j(w)|^2}{a_j(w)}$ *is called the norm of* ψ.

Definition. $\omega_F = \frac{\sqrt{-1}}{2\pi} \partial\bar{\partial} \log a_j(w)$ *represents the Chern class* $c(F)$ *of* F.

Definition.

$$u_\psi(z) = \frac{1}{4\pi} \log \frac{1}{|\psi|^2(f(z))}.$$

The divisor $(f^*\psi)$, the pull-back of (ψ) by f, is defined as the divisor associated to the system $\{\psi_j(f(z))\}$. The notation below will be used as in Sect. 6.

Notation

$$\begin{cases} \Delta_r = \{z \in \mathbb{C}^n \mid |z| < r\}; \quad |z|^2 = |z_1|^2 + \cdots |z_n|^2 \\ \partial\Delta_r = \{z \in \mathbb{C}^n \mid |z| = r\}, \\ dV(z) = \left(\dfrac{\sqrt{-1}}{2}\right)^n dz_1 \wedge d\bar{z}_1 \wedge \cdots \wedge dz_n \wedge d\bar{z}_n, \\ \sigma_\alpha(z) = dV(z_1, \ldots, z_{\alpha-1}, z_{\alpha+1}, \ldots, z_n), \\ \sigma(z) = \sum_{\alpha=1}^n \sigma_\alpha(z). \end{cases}$$

Definition.

$$\mathcal{N}(t, \psi) = \int_{(f^*\psi)\cap\Delta_t} \sigma(z).$$

Remark. $\mathcal{N}(t, \psi)$ amounts to the volume of $(f^*\psi) \cap \Delta_t$. More precisely, letting

$$(f^*\psi) = \sum_\nu m_\nu C_\nu,$$

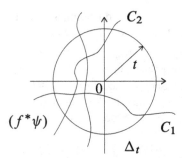

for $(n-1)$ dimensional irreducible analytic subsets C_ν, $\mathcal{N}(t, \psi)$ is defined as

$$\int_{(f^*\psi)\cap\Delta_t} \sigma(z) = \sum_\nu m_\nu \int_{C_\nu\cap\Delta_t\backslash(singular\ points)} \sigma(z).$$

Definition. $N(r, \psi) = \int_0^r \frac{n(t,\psi)}{S(t)} dt$, *where* $S(t)$ *stands for the volume of* $\partial\Delta_t$, *i.e.*

$$S(t) = \frac{2\pi^n}{(n-1)!} t^{2n-1}.$$

Definition.

$$\begin{cases} A(t) = \displaystyle\int_{\Delta_t} f^*\omega_F \wedge \sigma \\ T(r) = \displaystyle\int_0^r \frac{A(t)}{S(t)}\, dt \quad (= T_F(r)). \end{cases}$$

The average of a function g over $\partial \Delta_r$ is defined by

$$\mathcal{M}_r(g) = \frac{1}{S(t)} \int_{\partial \Delta_r} g(z)\, dS(z),$$

where $dS(z)$ is the volume element of $\partial \Delta_r$.

Definition. $m(r, \psi) = \mathcal{M}_r(u_\psi)$.

(2.0) $$dd^\perp u_\psi = f^*\omega_F.$$

Proof.

$$dd^\perp u_\psi = \frac{\sqrt{-1}}{2\pi} \partial\bar\partial \log \frac{1}{|\psi|^2} = \frac{\sqrt{-1}}{2\pi} \log a_j(f(z)) = f^*\omega_F.$$

Similarly to the case of one variable, the following holds for $T(r)$, $N(r, \psi)$ and $m(r, \psi)$.

Theorem. (2.1) **(The first main theorem)** *If* $f(0) \notin (\psi)$,

$$T(r) = N(r, \psi) + m(r, \psi) - m(0, \psi).$$

Proof. We put

$$\tau(t) = \frac{(n-2)!}{4\pi^n t^{2n-2}} \quad \left(d\tau = \frac{dt}{S(t)}\right).$$

Then

$$\begin{aligned} T(r) &= \int_0^r d\tau(t) \int_{\Delta_t} f^*\omega_F \wedge \sigma \\ &= \iint_{|z|<t<r} d(\tau(t)dd^\perp u_\psi \wedge \sigma(z)) = \int_{\Delta_r} [\tau(r) - \tau(|z|)]dd^\perp u_\psi(z) \wedge \sigma(z) \\ &= \sum_{\alpha_1}^n \int_{\Delta_r} [\tau(r) - \tau(|z|)]d_\alpha d^\perp u_\psi \wedge \sigma_\alpha. \end{aligned}$$

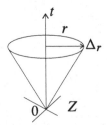

Here $d^\perp = \sqrt{-1}(\bar{\partial} - \partial)$ and d_α (resp.$d_\alpha^\perp$) stands for d (resp.$d^\perp$) with respect to z_α.

By choosing the coordinate z in advance, we may assume that $|\psi|^2(f(z_1, \ldots, z_\alpha, \ldots, z_n))$ is not identically zero in each $z\alpha$ when the other variables are fixed.

Let $z_\alpha'' = (z_1, \ldots, z_{\alpha-1}, z_{\alpha+1}, \ldots, z_n)$.

Then

$$T(r) = \sum_{\alpha=1}^{n} \lim_{\varepsilon \to 0} \int_{|z_\alpha''|<r} \sigma_\alpha \int_{[\Delta_r \backslash \Delta_\varepsilon]_{z_\alpha''}} [\tau(r) - \tau(|z|)]d_\alpha d_\alpha^\perp u_\psi.$$

Here $[\Delta_r - \Delta_\varepsilon]_{z_\alpha''} = [\Delta_r - \Delta_\varepsilon] \cap \{z_\alpha'' = const.\}$.

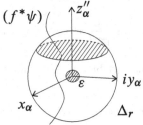

Since

$$[\tau(r) - \tau(|z|)]d_\alpha d_\alpha^\perp u_\psi = d_\alpha([\tau(r) - \tau(|z|)] d_\alpha^\perp u_\psi) + d_\alpha \tau(|z|) \wedge d_\alpha^\perp u_\psi$$
$$= d_\alpha([\tau(r) - \tau(|z|)]d_\alpha^\perp u_\psi) + d_\alpha u_\psi \wedge d_\alpha \tau(|z|)$$
$$(\text{by } d_\alpha \tau \wedge d_\alpha^\perp u_\psi = d_\alpha u_\psi \wedge d^\perp \tau),$$

we obtain

(2.2)
$$T(r) = \sum_{\alpha=1}^{n} \mathrm{I}_\alpha + \sum_{\alpha=1}^{n} \mathrm{II}_\alpha,$$

where

$$\mathrm{I}_\alpha = \lim_{\varepsilon \to 0} \int_{|z_\alpha''|<r} \sigma_\alpha \int_{[\Delta_r \backslash \Delta_\varepsilon]_{z_\alpha''}} d_\alpha([\tau(r) - \tau(|z|)]d_\alpha^\perp u_\psi)$$

and

$$\mathrm{II}_\alpha = \lim_{\varepsilon \to 0} \int_{|z''_\alpha| < r} \sigma_\alpha \int_{[\Delta_r \backslash \Delta_\varepsilon]_{z''_\alpha}} d_\alpha u_\psi \wedge d_\alpha^\perp \tau(|z|).$$

Let us compute I_1. We put $z''_1 = z''$, i.e. $z = (z_1, z'')$, $z'' = (z_2, \ldots, z_n)$. By fixing z'', let $\zeta_h(z'')$, $h = 1, 2, \ldots$ be the roots of $\psi_j(f(z_1, z'')) = 0$ (counted with multiplicity). For any a'' with $|a''| < r$, let us take r^0 so that $r^0 \neq |\zeta_h(a'')|$, $h = 1, 2, \cdots$. Then we consider $z = (z_1, z'')$ satisfying $|z_1| < r_0$ and $|z'' - a''| < \delta$. For such a z one has

$$|\psi|^2(f(z)) = \frac{|\psi_j(f(z))|^2}{a_j(f(z))} = \prod_{|\zeta_h(z'')| < r^0} |z_1 - \zeta_h(z'')|^2 U(z).$$

Here U is a C^∞ function with $U(z) \neq 0$.

By the definition of u_ψ,

(2.3) $$u_\psi(z) = \frac{1}{4\pi} \sum_h{}^{r_0} \log \frac{1}{|z_1 - \zeta_h(z'')|^2} + \{C^\infty \text{function}\},$$

where $\sum_h{}^{r_0}$ stands for the summation $\sum_{|\zeta_h(z'')| < r^0}$. Then, $u_\psi(z_1, z'')$ is continuous with respect to z'' for sufficiently small δ, since $\zeta_h(z'')$ then depends continuously on z''. Differentiating (2.3) one has

$$d_1 u_\psi(z) = -\frac{1}{4\pi} \sum_h{}^{r_0} \left(\frac{dz_1}{z_1 - \zeta_h} + \frac{d\bar{z}_1}{\bar{z}_1 - \bar{\zeta}_h} \right) + \{C^\infty \text{function}\}.$$

Therefore

$$d_1 u_\psi \wedge d_1^\perp \tau(|z|) = O\left(\frac{1}{|z|^{2n-2}} \sum_h{}^{r_0} \frac{1}{|z_1 - \zeta_h(z'')|} dz_1 \wedge d\bar{z}_1 \right).$$

Hence

$$\int_{[\Delta_r \backslash \Delta_\varepsilon]_{z''}} d_1 u_\psi \wedge d_1^\perp \tau(|z|)$$

converges uniformly with respect to z''. By Stokes's theorem,

$$\int_{[\Delta_r \backslash \Delta_\varepsilon]_{z''}} d_1 U_\psi \wedge d_1^\perp \tau(|z|)$$

$$= \int [\Delta_r \backslash \Delta_\varepsilon]_{z''} d_1 (u_\psi d_1^\perp \tau) - \int_{[\Delta_r \backslash \Delta_\varepsilon]_{z''}} u_\psi \wedge d_1 d_1^\perp \tau(|z|)$$

$$= \int_{[\Delta_r \backslash \Delta_\varepsilon]_{z''}} d_1 (u_\psi d_1^\perp \tau) = \int_{[\partial \Delta_r \backslash \partial \Delta_\varepsilon]_{z''}} u_\psi(z) d_1^\perp \tau(|z|).$$

By letting $z_1 = r_1 e^{i\theta_1}$,

$$d_1^\perp \tau(|z|) = \sqrt{-1}(\bar{\partial}_1 \tau - \partial_1 \tau) = \frac{(n-1)!}{4\pi^n} \frac{\sqrt{-1}}{|z|^{2n}} (z_1 \, d\bar{z}_1 - \bar{z}_1 \, dz_1)$$

$$= \frac{(n-1)!}{2\pi^n} \frac{1}{|z|^{2n}} r_1^2 \, d\theta_1 = \frac{1}{S(|z|)} \frac{r_1^2}{|z|} d\theta_1.$$

Hence we have

$$\text{(2.4)} \qquad \int_{\partial \Delta_r} u_\psi \, d_1^\perp \tau(|z|) = \frac{1}{S(r)} \frac{r_1^2}{r} \int_0^{2\pi} u_\psi(r_1 e^{i\theta_1}, z'') \, d\theta_1.$$

By Gauss's formula

$$\text{(2.5)} \qquad \frac{1}{2\pi} \int_0^{2\pi} \log |r_1 e^{i\theta_1} - \zeta| \, d\theta_1 = \max \{\log r_1, \log |\zeta|\}$$

integration of (2.3) yields

$$\int_0^{2\pi} u_\psi(r_1 e_1^{i\theta}, z'') \, d\theta_1 = \sum_h^{r_0} \min \left\{ \log \frac{1}{r_1}, \log \frac{1}{|\zeta_h(z'')|} \right\} + \{C^\infty \text{ function}\}.$$

Let us choose the fiber metric of F in such a way that

$$\text{(2.6)} \qquad |\psi|^2(w) = \frac{|\psi_j(w)|^2}{a_j(w)} \le e^{-4}$$

holds. Then $u_\psi(z) \ge \frac{1}{\pi}$.

Combining this with the preceding equality, we obtain

$$2 \le \int_0^{2\pi} u_\psi(r_1 e^{i\theta_1}, z'') \, d\theta_1 \le k \log \frac{1}{r_1} + \{C^\infty \text{ function}\}.$$

Here k stands for the number of ζ_h satisfying $|\zeta_h(z'')| < r^0$.

Therefore the right hand side of (2.4) is continuous in z'', and II_1 absolutely converges uniformly.

Lemma. (2.7) $\sum_{\alpha=1}^{n} \mathrm{II} = m(r, \psi) - m(0, \psi)$.

Proof. By (2.4),

$$\int_{|z''|<r} dV(z'') \int_{\partial \Delta_r} u_\psi(z) \, d_1^\perp \tau(|z|)$$

$$= \frac{1}{S(r)} \int_{|z''|<r} dV(z'') \int_0^{2\pi} u_\psi(r_1 e^{i\theta_1}, z'') \frac{r_1^2}{r} \, d\theta_1$$

$$= \frac{1}{S(r)} \int_{\partial \Delta_r} u_\psi(z) \frac{r_1^2}{r^2} \, dS(z)$$

$$\left(dS(z) = \frac{r}{r_1} r_1 \, d\theta \, dV(z'') \right).$$

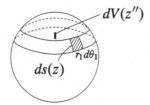

Since similar equalities hold for other α, we obtain

$$\sum_{\alpha=1}^{n} \mathrm{II} = \lim_{\varepsilon \to 0} \frac{1}{S(r)} \int_{\partial \Delta_r \setminus \partial \Delta_\varepsilon} u_\psi(z) \, dS(z) = m(r, \psi) - m(0, \psi)$$

by using $r^2 = r_1^2 + \cdots + r_n^2$.

(2.8) *Remark* The above proof shows also that $\mathcal{M}_r(u_\psi)$ converges absolutely.

(2.9)
$$\sum_{\alpha=1}^{n} \int_{\Delta_r \setminus \Delta_\varepsilon} d_\alpha d_\alpha^\perp \tau(|z|) \wedge \sigma_\alpha = 0.$$

Proof. The left hand side is

$$= \frac{(n-1)!}{\pi^n} \int_{\Delta_r \setminus \Delta_\varepsilon} |z|^{-2n-2} \sum_{\alpha=1}^{n} (|z|^2 - n|z_\alpha|^2) \, dV(z) = 0$$

(since $\sum_{\alpha=1}^{n} (|z|^2 - n|z_\alpha|^2) = n|z|^2 - n \sum_{\alpha=1}^{n} |z_\alpha|^2 = 0$).

(2.10) Lemma. $\sum_{\alpha=1}^{n} I_\alpha = N(r, \psi)$.

Proof. Since $r_1 = \sqrt{r^2 - |z''|^2}$, $[\Delta_\varepsilon]_{z''} = \varnothing$ if $|z''| \geq \varepsilon$. $\tau(r) - \tau(|z|) = 0$ if $|z_1| = r$.

By Stokes's theorem,

$$\int_{[\Delta_r \setminus \Delta_\varepsilon]_{z''}} d_1([\tau(r) - \tau(|z|)]d_1^\perp u_\psi)$$

$$= -\sum_h {}^{r_0} \oint [\tau(r) - \tau(|z|)]d_1^\perp u_\psi - [\tau(r) - \tau(\varepsilon)]\int_{[\partial\Delta_\varepsilon]_{z''}} d_1^\perp u_\psi$$

$$= \sum_h {}^{r_0}[\tau(r) - \tau(|(\zeta_h(z''), z'')|)] - [\tau(r) - \tau(\varepsilon)]\int_{[\Delta_\varepsilon]_{z''}} d_1 d_1 \perp u_\psi$$

(cf. the proof of Theorem (2.1)).

Hence, by integration

$$\int_{|z''|<r} \sigma_1 \int_{[\Delta_r \setminus \Delta_\varepsilon]_{z''}} d_1[\tau(r) - \tau(|z|)]d_1^\perp u_\psi$$

$$= \int_{|z''|<r} \sum_h {}^{r_0}[\tau(r) - \tau(|z|)]_{z=(\zeta_h(z''),z'')}\sigma_1 - [\tau(r) - \tau(\varepsilon)]\int_{\Delta_\varepsilon} d_1 d_1^\perp u_\psi \wedge \sigma_1.$$

The second term converges to 0 as $\varepsilon \to 0$ (because $\tau(\varepsilon) = O(\frac{1}{\varepsilon^{2n-2}})$ and $\int_{\Delta_\varepsilon} d_1 d_1^\perp u_\psi \wedge \sigma_1 = O(\varepsilon^{2n})$).

Since the map $z'' \to \sum_h (\zeta_h(z''), z'')$ is a parameter representation of $(f^*\psi)$,

$$I_1 = \int_{(f^*\psi)\cap\Delta_r} [\tau(r) - \tau(|z|)]\sigma_1.$$

Hence

$$\sum_{\alpha=1}^{n} I_\alpha = \sum_{\alpha=1}^{n} \int_{(f^*\psi)\cap\Delta_r} [\tau(r) - \tau(|z|)]\sigma_\alpha$$

$$= \int_{(f^*\psi)\cap\Delta_r} [\tau(r) - \tau(|z|)]\sigma = \int_{|z|<t<r,z\in(f^*\psi)} d\tau(t)\sigma$$

$$= \int_0^r d\tau(t) \int_{(f^*\psi)\cap\Delta_t} \sigma = \int_0^r d\tau(t)\,\mathcal{N}(t, \psi) = N(r, \psi).$$

Therefore, by combining (2.2) with Lemma (2.7) and Lemma (2.10), we obtain

$$T(r) = N(r, \psi) + m(r, \psi) - m(0, \psi)$$

so that the proof of Theorem (2.1) is completed.

(2.11) *Remark.* We shall write $T(r)$ as $T_F(r)$ to show the dependence of $T(r)$ on the bundle F. If $F = \{\hat{f}_{jk}\}$ for another cocyle $\hat{f}_{jk}$, there exists a system of holomorphic functions $\{h_j\}$ without zeros satisfying

$$\hat{f}_{jk} = \frac{h_k}{h_j} f_{jk}.$$

Putting $\hat{a}_j = a_j |h_j|^2$ one has $\hat{a}_j = |\hat{f}_{jk}|^2 \hat{a}_k$. Hence $\{\hat{a}_j\}$ is also a fiber metric of F.
  Setting $\hat{\omega}_F = \frac{\sqrt{-1}}{2\pi} \partial \bar{\partial} \log \hat{a}_j$, and

$$\hat{T}_F(r) = \int_0^r d\tau(t) \int_{\Delta_t} f^* \hat{\omega}_F \wedge \sigma,$$

one has

$$\hat{T}_F(r) - T_F(r) = \int_0^r d\tau(t) \int_{\Delta_t} \frac{1}{4\pi} dd^{\perp} \log a(f(z)).$$

Here we put $a = \frac{\hat{a}_j}{a_j}$. Hence, regarding the function a as a fiber metric of the trivial line bundle and applying the first main theorem (2.1) for $\psi = 1$, we obtain

$$\hat{T}_F(r) - T_F(r) = m(r, 1) - m(0, 1).$$

Since $m(r, 1) = \mathcal{M}_r(\frac{1}{4\pi} \log a(f(z)))$,

$$-\infty < \frac{1}{4\pi} \log \min_{w \in W} a(w) \leq m(r, 1) \leq \frac{1}{4\pi} \log \max_{w \in W} a(w) < +\infty.$$

  Therefore $T_F(r)$ is unique up to a bounded function in r. The same is true for the change of fiber metrics of F.

Proposition. (2.12) If $\omega_f > 0$ (cf. (6.5)),

$$\liminf_{r \to \infty} \frac{T_F(r)}{\log r} > 0.$$

For the proof we need the following two lemmas.

(2.13) **Lemma.** *Let* $\varphi = f^* \omega_F$. *Then*

$$\int_{\Delta_r} \varphi \wedge \sigma = \frac{r^{2n-2}}{2^{2n-2}(n-1)!} \int_{\Delta} \varphi \wedge (dd^{\perp} \log |z|^2)^{n-1}.$$

Proof. Since $d\varphi = 0$ on $\mathbb{C}^n$, there exists a 1-form λ satisfying $d\lambda = \varphi$, by virtue of de Rham's theorem. Furthermore

$$dd^{\perp} \log |z|^2 = \frac{2\sqrt{-1}}{|z|^2} \left(\sum_{\alpha=1}^{n} dz_{\alpha} \wedge d\bar{z}_{\alpha} - \frac{1}{|z|^2} \left(\sum_{\alpha=1}^{n} \bar{z}_{\alpha} dz_{\alpha} \right) \wedge \left(\sum_{\beta=1}^{n} z_{\beta} d\bar{z}_{\beta} \right) \right).$$

Hence

$$\int_{\Delta_r} d\lambda \wedge (dd^{\perp} \log |z|^2)^{n-1} = \int_{\partial \Delta_r} \lambda \wedge (dd^{\perp} \log |z|^2)^{n-1}$$

$$= \int_{\partial \Delta_r} \lambda \wedge \left(\frac{2\sqrt{-1}}{|z|^2} \right)^{n-1} \left(\sum_{\alpha=1}^{n} dz_{\alpha} \wedge d\bar{z}_{\alpha} \right)^{n-1}$$

(note that since $|z|^2 = r$ on $\partial \Delta_r$, $d|z|^2 = \sum_{\alpha=1}^{n} (\bar{z}_{\alpha} dz_{\alpha} + z_{\alpha} d\bar{z}_{\alpha}) = 0$)

$$= \frac{(n-1)! 2^{2n-2}}{r^{2n-2}} \int_{\partial \Delta_r} \lambda \wedge \sum_{\alpha=1}^{n} \sigma_{\alpha}$$

$$= \frac{(n-1)! 2^{2n-2}}{r^{2n-2}} \int_{\Delta_r} d\lambda \wedge \sigma = \frac{(n-1)! 2^{2n-2}}{r^{2n-2}} \int_{\Delta_r} \varphi \wedge \sigma.$$

Lemma. *Let* $f^* \omega_F = \frac{\sqrt{-1}}{2} \sum_{\alpha, \beta} g_{\alpha\beta} dz_{\alpha} \wedge d\bar{z}_{\beta}$. *Then*

$$\int_{\Delta_r} \sum_{\alpha=1}^{n} g_{\alpha\alpha}(z) \, dV(z) = r^{2n-2} \int_{\Delta_r} \frac{\sum_{\alpha, \beta} g_{\alpha\beta} z_{\alpha} \bar{z}_{\beta}}{|z|^{2n}} \, dV(z).$$

Proof. By the previous lemma,

$$T_F(r) = \frac{(n-1)!}{2\pi^n} \int_0^r \frac{dt}{t} \int_{\Delta_t} \frac{\sum_{\alpha, \beta} g_{\alpha\beta} z_{\alpha} \bar{z}_{\beta}}{|z|^{2n}} \, dV(z).$$

Since $\omega_F > 0$ by assumption, there exists a point where $f^* \omega_F < 0$ holds. (Recall that the Jacobian of f is not identically zero.) Hence

$$\int_{\Delta_t} \frac{\sum_{\alpha, \beta} g_{\alpha\beta} z_{\alpha} \bar{z}_{\beta}}{|z|^{2n}} \, dV(z) > 0.$$

so that we obtain the desired conclusion

$$\liminf_{r \to \infty} \frac{T_F(r)}{\log r} > 0.$$

Let us fix the local expressions of $f : \mathbb{C}^n \to W$ as

$$f : z = (z_1, \ldots, z_n) \to (w_j^1, \ldots, w_j^n) = (f_j^1(z), \ldots, f_j^n(z)),$$

with respect to a covering $W = \bigcup_j U_j$.

Definition.

$$J_j(z) = \det \begin{pmatrix} \dfrac{\partial w_j^1}{\partial z_1} & \cdots & \dfrac{\partial w_j^n}{\partial z_1} \\ \cdot & \cdot & \cdot \\ \cdot & \cdot & \cdot \\ \dfrac{\partial w_j^1}{\partial z_n} & \cdots & \dfrac{\partial w_j^n}{\partial z_n} \end{pmatrix}$$

is holomorphic on $f^{-1}(U_j)$ and defines a divisor $(J) = (J_j) = (J_k)$, since

$$\frac{J_j(z)}{J_k(z)} \neq 0 \quad \text{on} \quad f^{-1}(U_j) \cap f^{-1}(U_k).$$

Definition.

$$\begin{cases} N_1(t) = \displaystyle\int_{(J) \cap \Delta_t} \sigma, \\ N_1(r) = \displaystyle\int_0^r \frac{n_1(t)}{S(t)} \, dt. \end{cases}$$

Let v be a C^∞ volume form on W and let $v = b_j(w) \, dV(w_j^1, \ldots, w_j^n)$ on U_j.

Definition. $\omega_K = \frac{\sqrt{-1}}{2\pi} \partial \bar{\partial} \log b_j(w)$.
The $(1,1)$-form ω_K represents $c(L) = -c_1$, where K denotes the canonical bundle (7.3).
Let a $C^i nfty$ function ξ on $\mathbb{C}^n$ be defined by

$$f^* v = \xi(z) \, dV(z).$$

Then

$$\xi(z) = b_j(f(z)) |J_j(z)|^2 \quad \text{on} \quad f^{-1}(U_j).$$

Definition. $M(r) = \mathcal{M}_r(\frac{1}{4\pi} \log \xi)$.
The proof of the first main theorem (2.1) works to show the following.

Theorem. (2.14) *If $J(0) \neq 0$,*

$$T_K(r) = -N_1(r) + M(r) - M(0).$$

3   The Second Main Theorem

Let F be a line bundle over W and let $\psi_\lambda \in H^0(W, \mathcal{O}(F))$, $\lambda = 1, \dots, q$, be holomorphic sections. The following will be assumed below.

(3.1) *Assumption.*

  (i) $\omega_F > 0$.
  (ii) The divisors (ψ_λ) are nonsingular.
  (iii) The singularities of the divisor $\sum_{\lambda=1}^q (\psi_\lambda)$ are at most of normal crossings, i.e.
        $\psi_{\lambda_j}(w) = w_j^1 \cdots w_j^p$, $p \leq n$, with respect to some covering $W = \bigcup_j U_j$
        with local coordinates $(w_j^1, \dots, w_j^n)$ on U_j.

Let $f : \mathbb{C}^n \to W$ be a holomorphic map.

(3.2) *Assumption.* (i) $J(0) \neq 0$ (ii) $f(0) \neq (\psi_\lambda)$, $\lambda = 1, \dots, q$.

Our purpose here is to prove the following.

Theorem. (3.3) (**The second main theorem**) *Under the above assumptions,*

$$\sum_{\lambda=1}^q m(r, \psi_\lambda) + N_1(r) \leq -T_K(r) + O(\log T_F(r))$$

holds for $r \to +\infty$ *with* $r \notin E$, *for some open set* $E \subset \{r \geq 0\}$ *satisfying*

$$\int_E d(r^\beta) <; \infty \quad \text{for some } 0 < \beta < 1.$$

The proof will be done by the method of Sect. 11 extended to n variables. First we shall prepare four lemmas.

Definition. *For any* $\psi \in H^0(W, \mathcal{O}(F)) \setminus \{0\}$ *and for any constant* $\kappa > 0$, *we put*

$$\rho_\psi(w) = \frac{\kappa}{[\log |\psi|^2(w)]^2 |\psi|^2(w)}.$$

Definition. $\tilde{u}_\psi(z) = \frac{1}{4\pi} \log \rho_\psi(f(z)).$

Definition.
$$\mathcal{L}_\psi(z) = \frac{1}{2\pi} \log \log \frac{1}{|\psi|^2(f(z))}.$$

For the above constant $\kappa > 0$ we set $\kappa_1 = \frac{1}{4\pi} \log \kappa$ and

(3.4) $\tilde{u}_\psi(z) = u_\psi(z) - \mathcal{L}_\psi(z) + \kappa_1.$

The following notation will be used.

Notation

$$\begin{cases} \tilde{A}_\psi(t) = \int_{\Delta_t} dd^\perp \tilde{u}_\psi(z) \wedge \sigma, \\ \tilde{T}_\psi(r) = \int_0^r \tilde{A}_\psi(t) \, d\tau(t), \\ R_\psi(r) = \int_0^r d\tau(t) \int_{\Delta_t} dd^\perp \mathcal{L}_\psi(z) \wedge \sigma, \end{cases}$$

where $d\tau(t) = \frac{dt}{S(t)}$.

Combining (3.4) and (2.0) one has

(3.5) $$\tilde{T}_\psi(r) = T_F(r) - R_\psi(r) + const.$$

As a fiber metric of F we choose one satisfying $|\psi|^2(w) \le e^{-4}$. Then

(3.6) $$\log |\psi|^2 \ge \frac{1}{4}.$$

Lemma. *The integral*

$$\int_{\Delta_t} dd^\perp \mathcal{L}_\psi(z) \wedge \sigma$$

is absolutely convergent.

Proof. Since

$$dd^\perp \mathcal{L}_\psi \wedge \sigma = \sum_{\alpha=1}^n d_\alpha d_\alpha^\perp \mathcal{L}_\psi \wedge \sigma_\alpha,$$

it suffices to show that

$$\int_{\Delta_t} d_1 d_1^\perp \mathcal{L}_\psi \wedge \sigma_1$$

converges absolutely. Since $\mathcal{L}_\psi$ is C^∞ outside $(f^*\psi)$, we need to show that, for any point $c = (c_1, c'') \in (f^*\psi)$ and

$$\mathcal{U} = \{(z, z'') \in \mathbb{C}^n \mid |z_1 - c_1| < \varepsilon, |z'' - c''| < \varepsilon\},$$

$\int_{\mathcal{U}} d_1 d_1^\perp \mathcal{L}_\psi \wedge \sigma_1$ converges absolutely.

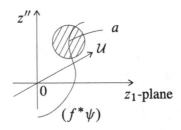

Since

$$\int_{\mathcal{U}} d_1 d_1^{\perp} \mathcal{L}_{\psi} \wedge \sigma_1 = \int_{|z''-c''|<\varepsilon} dV(z'') \int_{|z_1-c_1|<\varepsilon} d_1 d_1^{\perp} \mathcal{L}_{\psi},$$

it suffices to show that $\int_{|z_1-c_1|<\varepsilon} d_1 d_1^{\perp} \mathcal{L}_{\psi}$ converges absolutely, uniformly in z'' on $|z''-c''| < \varepsilon$. By a direct computation one has

$$d_1 d_1^{\perp} \mathcal{L}_{\psi} = \frac{2}{\log \frac{1}{|\psi|^2}} \frac{1}{4\pi} d_1 d_1^{\perp} \log \frac{1}{|\psi|^2} - \frac{1}{[\log \frac{1}{|\psi|^2}]^2} \frac{\sqrt{-1}}{\pi} \partial_1 \log 1 |\psi|^2 \wedge \bar{\partial}_1 \log \frac{1}{|\psi|^2}.$$

The first term is bounded, so that no problem is there as for the convergence. Let us write the second term as $G(z_1, z'')$ and estimate $\int_{|z_1-c_1|<\varepsilon} G(z, z'')$ by assuming $f(\mathcal{U}) \subset U_j$.

Setting $\ell = \log a_j(f(z))$ we have

$$\partial_1 \log \frac{1}{|\psi|^2} = -\frac{1}{\psi_j} \frac{\partial \psi_j}{\partial z_1} + \frac{\partial \ell}{\partial z_1},$$

since $|\psi|^2 = \frac{|\psi_j|^2}{a_j}$ by definition.

By choosing $a_j \geq 1$ in advance, we may assume that

$$G(z_1, z'') = \frac{1}{[\log \frac{1}{|\psi|^2}]^2} \left| \frac{1}{\psi_j} \frac{\partial \psi_j}{\partial z_1} - \frac{\partial \ell}{\partial z_1} \right|^2 \frac{\sqrt{-1}}{\pi} dz_1 \wedge d\bar{z}_1 \leq A + B,$$

where

$$A = \frac{2}{[\log |\psi_j|^2]^2 |\psi_j|^2} \left| \frac{\partial \psi_j}{\partial z_1} \right|^2 \frac{\sqrt{-1}}{\pi} dz_1 \wedge d\bar{z}_1$$

and

$$B = \frac{1}{[\log |\psi|^2]^2} \left| \frac{\partial \ell}{\partial z_1} \right|.$$

Since $|\frac{\partial \ell}{\partial z_1}|^2$ is C^∞ and $\frac{1}{[\log |\psi|^2]^2} \leq \frac{1}{2}$ by (3.6), B is bounded. A is written as

$$A = g^*\gamma, \quad \gamma = \frac{2\sqrt{-1}\,dw \wedge d\overline{w}}{\pi[\log|w|^2]^2|w|^2}$$

by using $g : z_1 \to w = \psi_j(z_1, z'')$.

Since

$$\int_{|w|<\varepsilon_0} \gamma = \frac{1}{\log\frac{1}{\varepsilon_0}} < +\infty,$$

the integral

$$\int_{|z_1-c_1|<\varepsilon} A = \int_{|z_1-a_1|<\varepsilon} g^*\gamma$$

converges absolutely.

Definition. $\mu(r, \psi) = \mathcal{M}_r(\mathcal{L}_\psi)$.

Lemma. $R_\psi(r) = \mu(r, \psi) - \mu(0, \psi)$.

Proof. By (3.6) one has

$$(3.7) \qquad 0 < \frac{1}{2\pi}\log 4 \leq \mathcal{L}_\psi = \frac{1}{2\pi}\log u_\psi + \frac{1}{2\pi}\log 4\pi \leq \frac{1}{2\pi}\log u_\psi + const.$$

Since $\mathcal{M}_r(u_\psi)$ converges absolutely (cf. Remark (2.8)), so does $\mu(r, \psi)$. Next we shall prove the following assertion:

$$(3.8) \qquad R_\psi(r) = \int_{\Delta_r} [\tau(r) - \tau(|z|)]dd^\perp\mathcal{L}_\psi \wedge \sigma.$$

Proof of (3.8): $dd^\perp\mathcal{L}_\psi$ is ∞ along the divisor $(f^*\psi)$. So we consider the ε-neighborhood $\mathcal{U}_\varepsilon$ and the 2ε-neighborhood $\mathcal{U}_{2\varepsilon}$ of $(f^*\psi)$ and choose a C^∞ function λ_ε such that

$$\lambda_\varepsilon(z) = \begin{cases} 0 & z \in \mathcal{U}_\varepsilon \\ 1 & z \in \mathcal{U}_{2\varepsilon}. \end{cases}$$

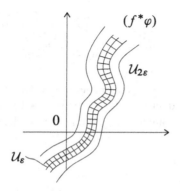

Then

$$R_\psi(r) = \lim_{\varepsilon \to 0} \iint_{|z|<t<r} d\tau \wedge \lambda_\varepsilon(z) dd^\perp \mathcal{L}_\psi \wedge \sigma$$

$$= \lim_{\varepsilon \to 0} \iint d(\tau \wedge \lambda_\varepsilon dd^\perp \mathcal{L}_\psi \wedge \sigma) - \lim_{\varepsilon \to 0} \iint \tau d\lambda_\varepsilon \wedge dd^\perp \mathcal{L}_\psi \wedge \sigma.$$

The second term is zero since $dd^\perp \mathcal{L}_\psi \wedge \sigma$ is already a $2n$-form. Hence, by applying Stokes's theorem to the first term we obtain

$$R_\psi(r) = \lim_{\varepsilon \to 0} \int \Delta_r[\tau(r) - \tau(|z|)]\lambda_\varepsilon(z) dd^\perp \mathcal{L}_\psi \wedge \sigma$$

$$= \int_{\Delta_r} [\tau(r) - \tau(|z|)] dd^\perp \mathcal{L}_\psi \wedge \sigma.$$

Continuation of the proof of Lemma: From (3.8), one can deduce immediately that

$$R_\psi(r) = N + \mu(r, \psi) - \mu(0, \psi),$$

where $N = \sum_{\alpha=1}^n I_\alpha$,

$$I_\alpha = \lim_{\varepsilon \to 0} \int_{|z_\alpha''|<r} \sigma_\alpha \int_{[\Delta_r \backslash \Delta_\varepsilon]_{z_\alpha''}} d_\alpha([\tau(r) - \tau(|z|)] d_\alpha^\perp \mathcal{L}_\psi)$$

by applying the method of the proof of the first main theorem (2.1).

Let us estimate I_1 when $\alpha = 1$. As before, let $\zeta_h(z'')$, $h = 1, 2, \ldots$ be the roots of $\psi(z_1, z'') = 0$. Then

$$\int_{[\Delta_r \backslash \Delta_\varepsilon]_\varepsilon} d_1([\tau(r) - \tau(|z|) d_1^\perp \mathcal{L}_\psi])$$

$$= -\sum_h \oint_{\zeta_h} d_1^\perp \mathcal{L}_\psi - [\tau(r) - \tau(\varepsilon)] \int_{[\Delta_\varepsilon]_{z''}} d_1 d_1^\perp \mathcal{L}_\psi.$$

Note that

$$d_1^\perp \mathcal{L}_\psi = \frac{2}{\log \frac{1}{|\psi|^2}} d_1^\perp u_\psi.$$

Since $|\psi|^2(f(z))$ is zero at $z = \zeta_h$, the first term on the right hand side of the integral is zero. Further, the second term tends to zero as $\varepsilon \to 0$, as in the proof of Lemma (2.10). Therefore $I_1 = 0$. Similarly $I_\alpha = 0$ for other α. Hence $N = 0$.

Notation

$$\rho(w) = \prod_{\lambda=1}^{q} \rho_{\psi_\lambda}(w).$$

(3.9) Lemma.
$\frac{1}{\rho(w)}(\frac{1}{4\pi}dd^\perp \log \rho(w))^n)$ *is continuous and positive on* W.

Proof.

$$\frac{1}{4\pi}dd^\perp \log \rho(w)$$

$$= \sum_{\lambda=1}^{q}\left(1 - \frac{2}{\log\frac{1}{|\psi|^2}}\right)\omega_F + \sum_{\lambda=1}^{q}\frac{1}{[\log|\psi_\lambda|^2]^2}\frac{\sqrt{-1}}{\pi}\partial\log\frac{1}{|\psi_\lambda|^2}\wedge\bar\partial\log\frac{1}{|\psi_\lambda|^2}$$

$$= \alpha\omega_F + \sum_{\lambda=1}^{q}\gamma_\lambda.$$

Here

$$\begin{cases} \alpha = \displaystyle\sum_{\lambda=1}^{q}\left(1 - \frac{2}{\log\frac{1}{|\psi_\lambda|^2}}\right) \text{ is a } C^\infty \text{ function} \\ \gamma_\lambda = \dfrac{1}{[\log|\psi_\lambda|^2]^2}\dfrac{\sqrt{-1}}{\pi}\partial\log\dfrac{1}{|\psi_\lambda|^2}\wedge\bar\partial\log\dfrac{1}{|\psi_\lambda|^2}. \end{cases}$$

In particular $\gamma_\lambda \wedge \gamma_\lambda = 0$. For simplicity we set $\omega = \omega_F$. Then

$$\left(\frac{1}{4\pi}dd^\perp \log \rho(w)\right)^n = \sum_{k=0}^{n} {}_nC_k(\alpha\omega)^{n-k}\wedge\left(\sum_{\lambda=1}^{q}\gamma_\lambda\right)^k$$

$$= \sum_{k=0}^{n} {}_nC_k(\alpha\omega)^{n-k}\wedge k!\sum_{\lambda_1<\cdots<\lambda_k}\gamma_{\lambda_1}\wedge\cdots\wedge\gamma_{\lambda_k}.$$

Simplifying the notation as $\rho_\lambda = \rho_{\psi_\lambda}$ and setting $\{\lambda_1, \ldots, \lambda_k, \nu_1, \ldots, \nu_{q-k}\} = \{1, \ldots, 1\}$ with $\nu_1, \ldots, \nu_{q-k}$, one has

$$\Omega = \frac{1}{\rho}\left(\frac{1}{4\pi}dd^\perp \log \rho(w)\right)^n$$

$$= \sum_{k=0}^{n} {}_nC_k(\alpha\omega)^{n-k}\wedge k!\sum_{\lambda_1<\cdots<\lambda_k}\frac{1}{\rho_{\nu_1}\cdots\rho_{\nu_{q-k}}}\frac{\gamma_{\lambda_1}}{\rho_{\lambda_1}}\wedge\cdots\wedge\frac{\gamma_{\lambda_k}}{\rho_{\lambda_k}}.$$

Note that $\frac{1}{\rho_\lambda} = \frac{1}{\kappa}[\log|\psi_\lambda|^2]^2|\psi_\lambda|^2$ are continuous on W. Moreover

$$\partial \log \frac{1}{|\psi_\lambda|^2} = -\frac{d\psi_{\lambda j}}{\psi_{\lambda j}} + \partial \log a_{\lambda j}$$

holds on U_j.

Hence

(3.10) $$\frac{\gamma_\lambda}{\rho_\gamma} = \frac{\sqrt{-1}}{\pi \kappa} \frac{1}{a_{\lambda j}} (d\psi_{\lambda j} - \psi_{\lambda j} \partial \log a_{\lambda j}) \wedge (d\overline{\psi}_{\lambda j} - \overline{\psi}_{\lambda j} \overline{\partial} \log a_{\lambda j}).$$

Therefore $\frac{\gamma_\lambda}{\rho_\lambda}$ is continuous on W, so that Ω is continuous on W.
Now, since $\gamma_\lambda \geq 0$ and

$$\alpha = \sum_{\lambda=1}^{q} \left(1 - \frac{2}{\log \frac{1}{|\psi_\lambda|^2}}\right) \geq \frac{q}{2} > 0 \quad \text{(cf. (14.6))},$$

$\Omega > 0$ holds where $\rho(w) \neq 0$. Hence it remains to show that $\Omega > 0$ at $w_0 \in \sum_{\lambda=1}^{q} (\psi_\lambda)$. Let $w_0 \in U_j$ and

$$\begin{cases} w_0 \in (\psi_\lambda) & \lambda = 1, \ldots, p \\ w_0 \notin (\psi_\lambda) & \lambda \geq p+1. \end{cases}$$

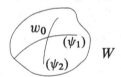

Then, by the condition (3.1) one can find a local coordinate $(w_j^1, \ldots, w_j^n)$ around w_0 such that $\psi_{\lambda j}(w) = w_j^\lambda$, $\lambda = 1, \ldots, p$.

Obviously,

$$\begin{cases} \dfrac{1}{4\pi} dd^\perp \log \rho(w) \geq \alpha \omega + \sum_{\lambda=1}^{p} \gamma_\lambda \\ \dfrac{1}{\rho_\lambda(w_0)} = 0, \quad \lambda = 1, \ldots, p. \end{cases}$$

so that we obtain

$$\Omega \geq {}_nC_p(\alpha\omega)^{n-p} p! \wedge \frac{1}{\rho_{p+1} \cdots \rho_q} \frac{\gamma_1}{\rho_1} \wedge \cdots \wedge \frac{\gamma_p}{\rho_p}.$$

If $1 \leq \lambda \leq p$,

$$\left[\frac{\gamma_\lambda}{\rho_\lambda}\right]_{w=w_0} = \frac{\sqrt{-1}}{\pi\kappa a_{\lambda j}(w_0)} dw_j^\lambda \wedge d\overline{w}_j^\lambda.$$

Hence

$$\Omega \geq \frac{{}_nC_k p! \alpha^{n-p}}{(\pi\kappa)^p} \left(\frac{1}{a_{1j}\cdots a_{pj}\rho_{p+1}\cdots\rho_q}\right)_{w=w_0}$$
$$\cdot \omega^{n-p} \wedge (\sqrt{-1})^p dw_j^1 \wedge d\overline{w}_j^1 \wedge \cdots \wedge dw_j^n \wedge d\overline{w}_j^n > 0.$$

Let $v = b_j(w) dV(w_j)$ $(b_j(w) > 0)$ be a volume form on W. Then, similarly to the one-variable case (cf. Lemma (3.11)) one has

$$\int_W \rho(w)v < +\infty.$$

Hence

$$\int_{\delta_r} f^*(\rho v) = \int_{\Delta_r} \rho(f(z))\xi(z) dV(z) < +\infty.$$

(3.11) Definition.

$$\begin{cases} \Psi(r) = \displaystyle\int_{\Delta_r} (\rho(f(z))\xi(z))^{\frac{1}{n}}, \\[2mm] \Phi(t) = M_t((\rho(f)\xi)^{\frac{1}{n}}) = \dfrac{1}{S(t)} \displaystyle\int_{\partial\Delta_t} (\rho(f(z))\xi(z))^{\frac{1}{n}} dS(z), \\[2mm] \Xi(r) = \displaystyle\int_0^r \Psi(t)\dfrac{dt}{S(t)}. \end{cases}$$

It follows from the definition that

$$\Psi(r) = \int_0^r dt \int_{\partial\Delta_t} (\rho(f(z))\xi(z))^{\frac{1}{n}} dS(z) = \int_0^r \Phi(t)S(t) dt.$$

Remark. This shows the measurability of $\Phi(t)$ by Fubini's theorem.

The proof of the following is similar to the case of one variable.

(3.12) Lemma. *Let* $0 < \beta < 1$ *and put* $\nu = (\frac{4n-2}{2n-2+\beta} - 1)^2$. *Then* $\Phi(r) \leq \Xi(r)^r$ *holds on* $[0, \infty) \setminus E$, *where* E *satisfies* $\int_E d(r^\beta) < +\infty$.

Proof of the second main theorem. In view of theorem (2.14), it suffices to show that

$$\sum_{\lambda=1}^n m(r, \psi_\lambda) + M(r) \leq O(\log T(r))$$

holds for $T(r) = T_F(r)$.

By taking the average of (3.7) over $\partial\Delta_t$, one has

$$\frac{1}{2\pi} \log 4 \leq \mu(r, \psi_\lambda) \frac{1}{2\pi} m(r, \psi_\lambda) + \frac{1}{2\pi} \log 4\pi.$$

Averaging (3.4) yields

$$m(r, \psi_\lambda) = \tilde{m}(r, \psi_\lambda) + \mu(r, \psi_\lambda) - \kappa_1.$$

Hence, combining this with the first main Theorem (2.1), it suffices to show

$$\sum_{\lambda=1}^{q} \tilde{m}(r, \psi_\lambda) + M(r) \leq O(\log T(r))$$

to prove the second main theorem.

First we observe that

$$\sum_{\lambda=1}^{q} \tilde{m}(r, \psi_\lambda) + M(r) = \mathcal{M}_r \left(\sum_{\lambda=1}^{q} \tilde{u}_\psi + \frac{1}{4\pi} \log \xi \right)$$

$$= \frac{1}{4\pi} \mathcal{M}_r \left(\log \left(\prod_{\lambda=1}^{q} \rho_{\psi_\lambda} \cdot \xi \right) \right) = \frac{1}{4\pi} \mathcal{M}_r (\log (\rho(f)\xi))$$

$$= \frac{n}{4\pi} \mathcal{M}_r (\log (\rho(f)\xi)^{\frac{1}{n}}) \leq \frac{n}{4\pi} \log \mathcal{M}_r ((\rho(f)\xi)^{\frac{1}{n}})$$

$$= \frac{n}{4\pi} \log \Phi(r).$$

Hence

(3.13) $$\sum \tilde{m}(r, \psi_\lambda) + M(r) \leq \frac{n\nu}{4\pi} \log \Xi(r)$$

holds for any $r \notin E$, by Lemma (3.12).

By Lemma (3.9), for any volume form v one can find $\kappa > 0$ such that

$$\kappa^q v \leq \frac{\kappa^q}{\rho} \left(\frac{1}{4\pi} dd^\perp \log \rho(w) \right)^n.$$

Note that $\frac{\kappa^q}{\rho}$ and $dd^\perp \log \rho$ do not depend on κ. Hence we may assume that

$$\rho v \leq \left(\frac{1}{4\pi} dd^\perp \log \rho(w) \right)^n$$

and accordingly

$$f^*(\rho v) = \rho(f(z))\xi(z)\,dV(z) \le \left(\frac{1}{4\pi}dd^\perp \log \rho(f(z))\right)^n$$

hold, by replacing v by $\kappa^q v$ if necessary.

We set

$$\frac{1}{4\pi}dd^\perp \log \rho(f(z)) = \frac{\sqrt{-1}}{2}\sum_{\alpha,\beta} h_{\alpha\beta}\,dz_\alpha \wedge d\bar{z}_\beta.$$

Then $(h_{\alpha\beta})$ is positive semidefinite in view of the proof of Lemma (3.9).

Since

$$\left(\frac{\sqrt{-1}}{2}\sum_{\alpha,\beta} h_{\alpha\beta}\,dz_\alpha \wedge d\bar{z}_\beta\right)^n = n!\det(h_{\alpha\beta}(z))\,dV(z),$$

(3.14) implies

$$(\rho(f(z))\xi(z))^{\frac{1}{n}} \le n(\det(h_{\alpha\beta}))^{\frac{1}{n}} \le \sum_{\alpha=1}^n h_{\alpha\alpha}(z).$$

Therefore

(3.15)
$$(\rho(f(z))\xi(z))^{\frac{1}{n}}\,dV(z) \le \left(\sum_{\alpha=1}^n h_{\alpha\alpha}(z)\right)dV(z)$$

$$= \frac{1}{4\pi}dd^\perp \log \rho(f(z)) \wedge \sigma = \sum dd^\perp \tilde{u}_{\psi_\lambda} \wedge \sigma.$$

Integration over Δ_t yields

$$\Psi(t) \le \sum_{\lambda=1}^q \tilde{A}_{\psi_\lambda}(t).$$

Hence

$$\Xi(r) = \int_0^r \Psi(t)\frac{dt}{S(t)}$$

$$\le \sum_{\lambda=1}^q \int_0^r \tilde{A}_{\psi_\lambda}(t)\frac{dt}{S(t)} = \sum_{\lambda=1}^q \tilde{T}_{\psi_\lambda}(r) \le qT(r) + const.$$

($\tilde{T}_\psi(r) \le T(r) + const.$ by (3.5).)

Combining this inequality with (3.13) we obtain

$$\sum_{\lambda=1}^q \tilde{m}(r,\psi_\lambda) + M(r) \le \frac{n\nu}{4\pi}\log(qT(r) + const.).$$

By letting $r \to +\infty$ we obtain the theorem.

4 Defect Relation

The defect relation describes how the image of $f(\mathbb{C}^n)$ of f catches the divisor (ψ).

Definition.

$$
\begin{cases}
\delta(\psi) = \liminf_{r \to +\infty} \dfrac{m(r.\psi)}{T(r)} \\
\delta_1 = \liminf_{r \to +\infty} \dfrac{N_1(r)}{T(r)}.
\end{cases}
$$

Remark. $\delta(\psi)$ is called the **defect** of f with respect to the divisor (ψ). By Theorem (2.1),

$$
\delta(\psi) = 1 - \limsup_{r \to +\infty} \frac{N(r, \psi)}{T(r)},
$$

so that $\delta(\psi)$ detects the bound of $N(r, \psi)$ with respect to $T(r)$. In particular, $\delta(\psi) = 1$ if $f(\mathbb{C}^n) \cap (\psi) = \varnothing$.

By the second main Theorem (3.3),

$$
\sum_{\lambda=1}^{q} \frac{m(r, \psi_\lambda)}{T(r)} + \frac{N_1(r)}{T(r)} \leq -\frac{T_K(r)}{T(r)} + \frac{O(\log(T(r)))}{T(r)}
$$

holds for $r \to +\infty$ and $r \notin E$.

Since $T(r) \to +\infty$ as $r \to +\infty$ (cf. Proposition (2.12)), the following relation holds between the defects.

Theorem. (5.1) (Defect relation)

$$
\sum_{\lambda=1}^{q} \delta(\psi_\lambda) + \delta_1 \leq \limsup_{r \to +\infty} \left[-\frac{T_K(r)}{T(r)} \right].
$$

Corollary 1. *Let* $f : \mathbb{C}^n \to \mathbb{P}_n$ *be a holomorphic map, let F be the hyperplane section bundle H, and let (ψ_λ), $\lambda = 1, \ldots, q$ be hyperplanes in general position. Then*

$$
\sum_{\lambda=1}^{q} \delta(\psi_\lambda) + \delta_1 \leq n + 1.
$$

In particular, if $f(\mathbb{C}^n) \cap (\psi_\lambda) = \varnothing$ holds for $\lambda = 1, \ldots, q$, one has $q \leq n + 1$ since $\delta(\psi_\lambda) = 1$ for all λ in this case. As a result, the number of hyperplanes omitted by f does not exceed $n + 1$, which is Picard's theorem (cf. Corollary 1 of (5.2)) if $n = 1$.

Proof of Corollary 1. Since $K = -(n+1)H$, $-\frac{T_K(r)}{T(r)} = n + 1$ from which the conclusion follows immediately.

Corollary 2. *Let $W = \mathbb{P}_n$, $F = kH$ and let (ψ) be a nonsingular hypersurface of degree k. Then*

$$\delta(\psi) + \delta_1 \leq \frac{n+1}{k}.$$

In particular, $k \leq n + 1$ if $f(\mathbb{C}^n \cap (\psi)) = \varphi$. Hence one has $f(\mathbb{C}^n) \cap (\psi) \neq \varphi$ holds for any nonsingular hypersurface (ψ) of degree $\geq n + 2$.

We note that the assumption that (ψ) is nonsingular cannot be removed.

Example. *(F. Sakai)* Let $f : \mathbb{C}^2 \to \mathbb{P}_2$ be defined by

$$f : z = (z_1, z_2) \to (w_0 : w_1 : w_2) = (1 : z_1 : z_1^k + e^{z_2}).$$

Then it is clear that the Jacobian of f is not equal to 0. Let $\psi = w_0^{k-1} w_2 - w_1^k$. Then (ψ) is a curve of degree k and $(0 : 0 : 1)$ is its singular point. $f(\mathbb{C}^2) \cap (\psi) = \varnothing$ since $\psi(f(z)) = e^{z_2} \neq 0$.

Example. In Bieberbach's example (cf. Sect. 12), $f(\mathbb{C}^2) \cap (\psi) \neq \varnothing$ for any nonsingular curve of degree k if $k \geq n + 2$, although $f(\mathbb{C}^2)$ occupies only a relatively small portion of $\mathbb{C}^2$.

Corollary 3. *Let $f : \mathbb{C}^n \to \mathbb{P}_n$ be a holomorphic map, let $(\psi_\lambda ambda)$ be hypersurfaces of degree k_λ for $\lambda = 1, \ldots, q$ and assume that (3.1) and (3.2) are satisfied. Then*

$$\sum_{\lambda=1}^{q} k_\lambda \delta(\psi_\lambda) + \delta_1 \leq n + 1.$$

In particular, $f(\mathbb{C}^n) \cap (\sum_{\lambda=1}^{q} (\psi_\lambda)) \neq \varnothing$ holds if $\sum_{\lambda=1}^{q} k_\lambda \geq n + 2$.

Proof. Let $F_\lambda = k_\lambda H$. Then

$$\delta(\psi_\lambda) = \liminf_{r \to +\infty} \frac{m(r, \psi_\lambda)}{T_{F_\lambda}(r)} = \liminf_{r \to +\infty} \frac{m(r, \psi_\lambda)}{k_\lambda T_H(r)}.$$

Remark. By the second main theorem,

$$\sum_{\lambda=1}^{q} \frac{m(r, \psi_\lambda)}{T_H(r)} + \frac{N_1(r)}{T_H(r)} \leq -\frac{T_K(r)}{T_H(r)} + \frac{O(\log(\sum_{\lambda=1}^{q} k_\lambda) T_H(r))}{T_H(r)}.$$

Hence, by letting $r \to +\infty$, one has

$$\sum_{\lambda=1}^{q} k_\lambda \delta(\psi_\lambda) + \delta_1 \leq n + 1.$$

5 Applications

Let W be a complex manifold of dimension n and let $f : \mathbb{C}^n \to W$ be a holomorphic map whose Jacobian J satisfies $J(0) \neq 0$. Let K be the canonical bundle of W (cf. (7.3)) and let L be another line bundle over W. Consider the line bundle $F = mK - L (:= K^m \otimes L^{-1})$. Let $\{a_j(w)\}$ and $\{b_j(w)\}$ be the fiber metrics of L and K, respectively, and let $\{b_j^m(w)a_j^{-1}(w)\}$ be the associated fiber metric of F.

Let us assume that F admits a nonzero holomorphic section $\psi = (\psi_j) \in H^0(W, \mathcal{O}(F))$. Then, by the first main Theorem (2.1),

$$T_F(r) = mT_K(r) - T_L(r) = N(r, \psi) + m(R, \psi) - m(r, \psi).$$

Combining this equality with Theorem (2.14), one has

$$\frac{1}{m}T_L(r) + \frac{1}{m}N(r, \psi) + N_1(r) = \left\{ M(r) - \frac{1}{m}m(r, \psi) \right\} - \left\{ M(0) - \frac{1}{m}m(0, \psi) \right\}.$$

Here

$$M(r) - \frac{1}{m}m(r, \psi) = \frac{1}{4\pi}\mathcal{M}_r\left(\log \xi - \frac{1}{m}\log\frac{1}{|\psi|^2} \right)$$

$$= \frac{1}{4\pi}\mathcal{M}_r(\log \xi |\psi|^{\frac{2}{m}})$$

$$= \frac{1}{4\pi}\mathcal{M}_r\left(\log\left(b_j|J_j|^2 \frac{a_j^{\frac{1}{m}}}{b_j}|\psi_j|^{\frac{2}{m}} \right) \right).$$

$(\xi = b_j|J_j|^2$ and $|\psi|^2 = \frac{|\psi_j|^2}{b_j^m a_j^{-1}})$

(5.1) Notation

$$\begin{cases} \zeta(z) = a_j(f(z))^{\frac{1}{m}}|\psi_j(f(z))|^{\frac{2}{m}}|J_j(z)|^2 & \text{on } f^{-1}(U_j) \\ \tilde{M} = \frac{1}{4\pi}\mathcal{M}_r(\log \zeta(z)). \end{cases}$$

Theorem. (5.2)

$$\frac{1}{m}T_L(r) + \frac{1}{m}N(r, \psi) + N_1(r) = \tilde{M}(r) - M(0).$$

Definition. $p_m = \dim H^0(W, \mathcal{O}(mK))$. p_m is called the **m-genus** of W.

We shall assume $p_m > 0$ below. Namely we only consider those W such that $H^0(W, \mathcal{O}(mK)) \neq \{0\}$.

Let us put $L = 0$, $F = mK$ and $a_j = a = const.$ in Theorem (5.2). Then $\tilde{v} = a^{\frac{1}{m}} |\psi_j(w)|^{\frac{2}{m}} dV(w_j)$ is a volume form on W. By choosing a so that $\zeta(0) = 1$, the theorem implies

$$0 \le \frac{1}{m} N(r, \psi) + N_1(r) = \tilde{M}(r) = \frac{1}{4\pi} M(\log \zeta(z))$$

$$\le \frac{1}{4\pi} \log \mathcal{M}_r(\zeta(z)) = \frac{1}{4\pi} \log \frac{1}{S(r)} \int_{\partial \Delta_r} \zeta(z) \, dS(z).$$

Hence

$$\int_{\partial \Delta_r} \zeta(z) \, dS(z) \ge S(r)$$

a nd

$$\int_{\Delta_r} \zeta(z) \, dV(z) \ge \frac{\pi^n}{n!} r^{2n}.$$

Since $f^*\tilde{v} = \zeta(z) \, dV(z)$ by definition, we obtain:

Corollary 1. *If $p_m > 0$ for some m, then there exists a volume form $\tilde{v}$ on W such that, for any holomorphic map $f : \mathbb{C}^n \to W$ with $J(0) \ne 0$,*

$$\int_{\Delta_r} f^*\tilde{v} \ge \frac{\pi^n}{n!} r^{2n}$$

holds.

(5.3) *Remark.* Since $a^{\frac{1}{m}} |\psi_j(w)|^{\frac{2}{m}}$ is continuous, $v \ge c_0 \tilde{v}$ holds for an ordinary volume form v on W for some positive constant c_0. Hence we have the following.

Corollary 2. *Under the situation of Corollary 1, $\int_{\Delta_r} f^*v \ge cr^{2n}$ holds, where $c = c_0 \frac{\pi^n}{n!}$.*

Example. Let W be a complex torus $\mathbb{C}^n/G$ and let f be the covering map. Then $K = 0$ so that $k_m = 1$ for all m. It is clear that $\int_{\Delta_r} f^*v$ is of the same order as r^{2n}.

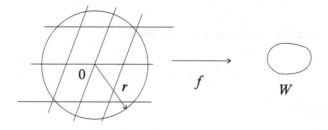

Definition.

$$\deg (f|\Delta_r) = \frac{\int_\Delta f^* v}{\int_W v}.$$

$\deg f|\Delta_r$ is called the **average mapping degree**.

In this terminology, Corollary 2 can be stated as follows.

Corollary 3. *If $p_m > 0$ for some m, then*

$$\liminf_{r \to +\infty} \frac{\deg (f|\Delta_r)}{r^{2n}} > 0.$$

Lemma. *If $(f^*\psi) \neq \varnothing$,*

$$\liminf_{r \to \infty} \frac{N(r, \psi)}{\log r} > 0.$$

Lemma. *If $(J) \neq 0$,*

$$\liminf \frac{N_1(r)}{\log r} > 0.$$

Similarly to the proof of Lemma (2.13),

$$\int_{(f^*\psi \cap \Delta_r)} \sigma = \frac{r^{2n-2}}{2^{2n-2}(n-1)!} \int_{(f^*\psi)\cap \Delta_r} (dd^\perp \log |z|^2)^{n-1} \geq Cr^{2n-2}.$$

Here C is a constant. Hence, noticing that $S(r) \sim r^{2n-1}$, we conclude that $N(r, \psi) \gtrsim \log r$ and $N_1(r) \gtrsim \log r$ as well.

By the above lemma, if $(f^*\psi) \neq 0$ or $(J) \neq 0$ there exists $\varepsilon > 0$ such that

$$\varepsilon \log r \leq \log \left\{ \frac{1}{S(r)} \int_{\partial \Delta_r} \zeta(z)\, dS(z) \right\}.$$

Therefore

$$\int_{\partial \Delta_r} \zeta(z)\, dS(z) \geq (const.) \times r^{2n-1+\varepsilon}.$$

Hence, by integration and Remark (5.3) we obtain

$$\int_\Delta f^* v \geq (const.) \times r^{2n+\varepsilon}.$$

and in other words:

Corollary 4. *Let $p_m > 0$ for some m. Then, for any f with $(f^*\psi) \neq 0$ or $(J) \neq 0$,*

$$\liminf_{r \to ;\infty} \frac{\deg (f|\Delta_r)}{r^{2n+\varepsilon}} > 0 \quad \text{for some } \varepsilon > 0.$$

Corollary 5. *If* $\deg (f|\Delta_r) = O(r^{2n}) \log r$, *then* $(J) = 0$, $(f^*\psi) = 0$ *and* $p_m \leq 1$ *for all* m.

Proof. Suppose that $p_m \geq 2$ for some m. Then there exist $\psi_\lambda \in H^0(W, \mathcal{O}(mK))$, $\lambda = 1, 2$ such that $f(\mathbb{C}^n \cap (\psi)) \neq \varnothing$ holds for some $\psi = a_1\psi_1 + a_2\psi_2$, $a_\lambda \in \mathbb{C}$.

Problem. (Hirzebruch[8]) Classify the compactifications of $\mathbb{C}^n$. (What are the compactifications of $\mathbb{C}^n$?)

Here a compact complex manifold W is said to be a compactification of $\mathbb{C}^n$ if there exists an analytic subset $a \subset W$ such that $W \setminus A = \mathbb{C}^n$.

$n = 1$: $\mathbb{P}_1$ is the unique compactification of $\mathbb{C}$.

$n = 2$: Compactifications of $\mathbb{C}^2$ are rational surfaces (a conjecture of Van de Ven [23]).

Theorem. (5.4) (cf. [15]) *Compactifications of* $\mathbb{C}^2$ *are rational.*

Proof. Let $W = \mathbb{C}^2 \cup A$ be a compactification of $\mathbb{C}^2$ and let $f : \mathbb{C}^2 \to \mathbb{C}^2 \hookrightarrow W$ be the embedding. Then it is obvious that $\deg (f|\Delta_r) < 1$. Hence $p_m = 0$ by Corollary 3. On the other hand, $b_1(W) \leq b_1(\mathbb{C}) = 0$ since $\operatorname{codim}_{\mathbb{R}} A \geq 2$. Therefore, by the classification theory of complex surfaces ([14] IV), W is rational.

Remark. The above argument carries over to some cases where A is not analytic. Morrow [17] gave an alternate proof of the theorem by determining A.

$n \geq 3$: Nothing is known more than $p_m = 0$.

We put $\mathbb{C}^* = \mathbb{C} \setminus \{0\}$.

Problem. What are the compactifications of $\mathbb{C} \times \mathbb{C}^*$?

Let $f : \mathbb{C}^2 \to \mathbb{C} \times \mathbb{C}^*$ be the holomorphic map $(z_1, z_2) \to (z_1, e^{z_2})$. Then $\deg (f|\Delta_r) = O(r)$. Hence

$$\liminf_{r \to +\infty} \frac{\deg (f|\Delta_r)}{r^2} = 0.$$

Therefore $p_m = 0$ by Corollary 3. Moreover, $b_1(W) \leq b_1(\mathbb{C}) \times \mathbb{C}^* \leq 1$ for any compactification W of $\mathbb{C} \times \mathbb{C}^*$. Hence W is either a rational surface or a surface with $b_1(W) = 1$. Compactifications with $b_1(W) = 1$ have not yet been classified (cf. Inoue [10]).

Next we consider holomorphic maps to algebraic manifolds of general type (cf. Sect. 7).

Theorem. (5.5) (cf. Corollary of Theorem (7.7)) *Let* W *be an algebraic manifold of general type. Then the Jacobian of any holomorphic map* $f : \mathbb{C}^n \to W$ *is identically zero.*

Proof. Let L be a hyperplane section bundle over W. Then, since W is of general type, there exist m and $\psi \in H^0(W, \mathcal{O}(mK - L))$ with $\psi \neq 0$ (cf. §7). Suppose that there exists a holomorphic map $f : \mathbb{C}^n \to W$ whose Jacobian J_f is not identically 0. Then

$$\frac{1}{m}T_L(r) \leq \tilde{M}(r) - \tilde{M}(0)$$

holds by Theorem (5.2).

Let

$$\tilde{v} = a_j(w)^{\frac{1}{m}}|\psi_j(w)|^{\frac{2}{m}} dV(w_j) \quad \text{(on } U_j)$$

be a volume form on W (cf. (7.5)). Since $\int_W \tilde{v} < +\infty$,

$$\int_{\Delta_r} f^*\tilde{v} = \int_{\Delta_r} \zeta(z) \, dV(z) < +\infty,$$

where $\zeta(z)$ is as in the notation (5.1).

Notation

$$\begin{cases} \Psi(r) = \displaystyle\int_{\Delta_r} \zeta^{\frac{1}{n}} \, dV(z), \\[2mm] \Phi(z) = \mathcal{M}_t(\zeta^{\frac{1}{n}}), \\[2mm] \Xi(r) = \displaystyle\int_0^r \frac{\Psi(t)}{S(t)} \, dt. \end{cases}$$

Similarly as in Lemma (3.12), one has the following.

Lemma. *Let* $\nu = \left(\frac{4n-2}{2n-2+\beta} - 1\right)^2$ *for* $0 < \beta < 1$. *Then there exists a subset* E *of* $\{r \geq 0\}$ *with* $\int_E d(r^\beta) < +\infty$ *such that* $\Phi(r) \leq \Xi(r)^\nu$ *holds outside* E.

Continuation of the proof of Theorem (5.5). Note that

$$\tilde{M}(r) = \frac{n}{4\pi}\mathcal{M}_r(\log \zeta^{\frac{1}{n}}) \leq \frac{n}{4\pi} \log \mathcal{M}_r(\zeta^{\frac{1}{n}})$$

$$= \frac{n}{4\pi} \log \Phi(r) \leq \frac{n\nu}{4\pi} \log \Xi(r).$$

Since L is the hyperplane section bundle,

$$\omega_L = \frac{\sqrt{-1}}{2\pi}\partial\bar{\partial} \log a_j(w) > 0$$

holds for some fiber metric $\{a_j(w)\}$ of L. Then $\tilde{v} \leq c_0\omega_L^n$ holds for some c_0. Since

$$f^*\tilde{v} \leq c_0(f^*\omega_L)^n,$$

similarly to (3.15) one has

$$\zeta(z)^{\frac{1}{n}} dV(z) \le c_0^{\frac{1}{n}} f^* \omega_L \wedge \sigma,$$

so that $\Xi(r) \le c_0 \frac{1}{n} T_L(r)$ follows by integration.

Combining the above inequalities we obtain

$$\frac{1}{m} T_L(r) \le \frac{n\nu}{4\pi} \log T_L(r) + const.$$

But this contradicts that $T_L(r) \to +\infty$ as $r \to +\infty$.

Problem. Find compact complex manifolds W whose universal covering space $\tilde{W}$ is $\mathbb{C}^n$.

From the above theorem $\tilde{W}$ cannot be of general type.

$n = 1$: W is a torus (an elliptic curve).

$n = 2$:

$$\begin{cases} \text{① tori;} \\ \text{② elliptic surfaces of the form } \mathbb{C}^n/G, \\ \quad \text{where } G \text{ is generated by affine transformations;} \\ \text{③ } b_1(W) = 1. \end{cases}$$

① and ② are completely understood (cf. Suwa [22], Kodaira [14] II).
③ is open.

Conjecture. Does $b_2(W) = 0$ hold if $\tilde{W} = \mathbb{C}^2$?

Problem. What is the Nevanlinna theory of $f : \mathbb{C} \to W$ for a compact complex manifold W of dimension n?

See [3, 24, 25] for the case $W = \mathbb{P}_n$.

References

1. Bieberbach, L.: Beispiel zweier Funktionen zweier komplexer Variablen welche eine schlicht volumetreue Abbildung des R_4 auf einer Teil seiner selbst vermitteln. Preuss. Akad. Wiss. Sitzungsbe. 476–479 (1933)
2. Bochner, S., Martin, W.T.: Several complex variables. Annals of Mathematics Studies, vol. 10. Princeton University Press, Princeton (1948)
3. Carlson, J.: Some degeneracy theorems for entire functions in an algebraic variety. Trans. Am. Math. Soc. **168**, 273–301 (1972)
4. Carlson, J., Griffiths, P.A.: A defect relation for equidimensional holomorphic mappings between algebraic varieties. Ann. Math. **95**, 557–584 (1972)
5. Cartan, H.: Les systèmes de fonctions holommorphes a variététes linéaires lacunaires leurs applications. Ann. Ec. Norm. **45**, 255–346 (1928)
6. Griffiths, P.: Holomorphic mappings into canonical algebraic varieties. Ann. Math. **93**, 439–458 (1971)
7. Griffiths, P., King, J.: Nevanlinna theory and holomorphic mappings between algebraic varieties. Acta Math. **130**, 145–220 (1973)
8. Hirzebruch, F.: Some problems on differentiable and complex manifolds. Ann. Math. **60**, 213–236 (1954)
9. Iitaka, S.: On algebraic varieties whose universal covering manifolds are complex affine 3-spaces I. In: Akizuki, Y. (ed.) Number Theory, Algebraic Geometry and Commutative Algebra, pp. 147–167. Kinokuniya, Tokyo (1973)
10. Inoue, M.: On surfaces of class VII_0. Proc. Jpn. Acad. **49**, 445–446 (1973)
11. Kobayashi, S.: Volume elements, holomorphic mappings and Schwarz's lemma. In: Entire functions and related topics, Proceedings of Symposia in Pure Mathematics /textbf11. La Jolla. California: Mmer Math. Soc. Provodence R. I, pp. 253-560 (1968)
12. Kobayashi, S.: Hyperbolic manifolds and holomorphic mappings. Marcel Dekker, New York (1970)
13. Kobayashi, S., Ochiai, T.: Mappings into compact complex manifolds with negative first Chern class. J. Math. Soc. Jpn. **23**, 137–148 (1971)
14. Kodaira, K.: On the structure of compact complex analytic surfaces, I, II, III, IV. Am. J. Math. **86**, 751-798 (1964). **88**, 682-721 (1966), **90**, 55-83 (1968), **90**, 1048–1066 (1968)
15. Kodaira, K.: On holomorphic mappings of polydiscs into compact complex manifolds. J. Differ. Geom. **6**, 33–46 (1971)
16. Lattès, M.S.: Sur les formes réduites des transformations ponctuelles à deux variables. C. R. **152**, 1566–1569 (1911)
17. Morrow, A.: Minimal compactifications of $\mathbb{C}^2$. Rice Univ. Stud. **59**, 97–112 (1973)
18. Nevanlinna, R.: Analytic Functions. Springer, New York (1970)
19. Shiffman, B.: Applications of geometric measure theory to value distribution theory for meromorphic maps (to appear)

© The Author(s) 2017

K. Kodaira, *Nevanlinna Theory*, SpringerBriefs in Mathematics,
https://doi.org/10.1007/978-981-10-6787-7

20. Siegel, C.L.: Meromorphe Funktionen auf kompakten analytishcen Mannigfaltigkeiten. Nachr. Akad. Wiss. Göttingen, 71–77 (1955)
21. Sternberg, S.: Local contractions and a theorem of Poincaré. Am. J. math. **79**, 809–824 (1957)
22. Suwa, T.: On hyperelliptic sufaces. J. Fac. Sci. Univ. Tokyo **16**, 469–473 (1970)
23. Van de Ven, A.: Analytic compactification of complex homology cells. Math. Ann. **147**, 189–204 (1962)
24. Weyl, H., Weyl, J.: Meromorphic functions and analytic curves. Annals of Mathematical Studies, vol. 12. Princeton University Press, Princeton (1943)
25. Wu, H.: The equidistribution theory of holomorphic curves. Annals of Mathematical Studies, vol. 64. Princeton University Press, Princeton (1970)

Printed in the United States
By Bookmasters